Water Resources Management in the People's Republic of China

Edited by Sun Xuetao, Robert Speed and Shen Dajun

I0125093

Routledge
Taylor & Francis Group

LONDON AND NEW YORK

First published 2010 by Routledge
2 Park Square, Milton Park, Abingdon, Oxon, OX14 4RN

Simultaneously published in the USA and Canada
by Routledge
52 Vanderbilt Avenue, New York, NY 10017

First issued in paperback 2013

Routledge is an imprint of the Taylor & Francis Group, an Informa business

British Library Cataloguing in Publication Data
A catalogue record for this book is available from the British Library

Typeset in Times by Value Chain, India

ISBN 13: 978-0-415-85203-6 (pbk)
ISBN 13: 978-0-415-54357-6 (hbk)

Water Resources Management in the People's Republic of China

Chinese water resource managers face a challenge that is both immense and unique. They must balance limited water supplies against the needs of the world's largest population; demands for rapid economic growth with calls for improved environmental management; and the desire for a market-based approach to the allocation of water with a history of State ownership and strict government control of all resources.

This book describes the development of a water rights system in the People's Republic of China. It covers different aspects of water resources management in China – including water planning, the provision of environmental flows, urban water management, and irrigation district management – and examines how these are being addressed through a rights-based approach. The book includes several detailed examples of the Chinese application of water rights as they address the diverse challenges of different basins across China.

This book previously appeared as a special issue of the *International Journal of Water Resources Development*.

Sun Xuetao is the Director-General of the Department of Water Resources, Ministry of Water Resources, People's Republic of China. In that role, he leads the formulation and implementation of China's policies for water resources management.

Robert Speed is a water policy consultant from Brisbane, Australia and was the Australian Team Leader on the China Water Entitlements and Trading project from 2006-8. Robert has held senior positions in the Water Reform Unit of the Queensland Department of Natural Resources and Water and was involved in the development and implementation of the Australian water sector reforms.

Shen Dajun is a senior engineer at the China Institute of Water Resources and Hydropower Research, Beijing, and has worked in water resources management for more than 10 years. He had been the team leader for more than 10 international and domestic water projects and has published more than 50 books and papers on water resources management.

Contents

Notes on Contributors   vii

1. Introduction: The Development of a Water Rights System in China   1
   Xuetao Sun

2. Water Resources Management in the People's Republic of China   5
   Bin Liu & Robert Speed

3. Water Resources Allocation in the People's Republic of China   21
   Dajun Shen & Robert Speed

4. Irrigation Development and Water Rights Reform in China   39
   Roger C. Calow, Simon E. Howarth & Jinxia Wang

5. Urban Water Management in China   61
   Martin Cosier & Dajun Shen

6. Transferring and Trading Water Rights in the People's Republic of China   81
   Robert Speed

7. Approaches to Providing and Managing Environmental Flows in China   95
   Xiqin Wang, Yuan Zhang & Cassandra James

8. An Asset-based, Holistic, Environmental Flows Assessment Approach   113
   Christopher J. Gippel, Nick R. Bond, Cassandra James & Xiqin Wang

9. Balancing Environmental Flows Needs and Water Supply Reliability   143
   Christopher J. Gippel, Martin Cosier, Sharmil Markar & Changshun Liu

10. A Harmonious Water Rights Allocation Model for Shiyang River Basin, Gansu Province, China   167
    Zhongjing Wang, Hang Zheng & Xuefeng Wang

11. A Water Rights Constitution for Hangjin Irrigation District, Inner Mongolia, China   185
    Hang Zheng, Zhongjing Wang, You Liang & Roger C. Calow

12. A Comparison of Water Rights Systems in China and Australia   201
      Robert Speed

      Index   218

Notes on Contributors

Nick R. Bond is a Research Fellow with the School of Biological Sciences, Monash University and eWater Cooperative Research Centre. Nick is a freshwater ecologist with research interests in the effects of natural and anthropogenic disturbance on streams, restoration ecology and flow ecology relationships. Nick has been involved in a number of projects to develop environmental flow recommendations for regulated and unregulated rivers.

Roger C. Calow is head of the Water Policy Programme at the UK's Overseas Development Institute (ODI) and works on international groundwater management issues for the British Geological Survey (BGS). In addition to his work on water rights and irrigation reform in Asia, he is currently leading research projects in Africa on sustainable service delivery, climate change and water security.

Martin Cosier has been working in China since early 2007. As a water policy specialist on the Water Entitlements and Trading Project, Martin provided policy advice on resource allocation planning, environmental flows and urban water supply issues. Prior to moving to China, Martin was a policy advisor with the Premier's Department in the Queensland Government. He has qualifications in law and environmental science.

Christopher J. Gippel is Director of Fluvial Systems Pty Ltd., a company that provides specialist consulting services in the areas of environmental flows, geomorphology, hydrology, river rehabilitation, and river and estuary management. He holds a BSc (Hons), University of Newcastle, and PhD, University of NSW.

Cassandra James is a Research Fellow at the Australian Rivers Institute, Griffith University. Her areas of expertise include environmental flows and plant ecology.

Bin Liu is the Division Director, Division of Water Resources Management, within the Department of Water Resources in China's Ministry of Water Resources.

Changshun Liu received a PhD degree from Beijing Normal University in 2004. He works as a Senior Engineer at the Water Development Research Centre within the Ministry of Water Resources, Beijing. His main research interests include water policy, water resources management models and technology.

Sharmil Markar is a director and Principal Engineer with WRM Water & Environment Pty Ltd., a water resources consultancy in Brisbane, Australia. He has over 25 years of experience in numerical modelling of hydrologic and hydraulic processes, as well as water resources management and planning, and flood forecasting, flood risk and damage assessment. He has worked in Australia, China, Indonesia, Iran, Vietnam and Sri Lanka.

Xiqin Wang is an Associate Professor at the School of Environment and Natural Resources, Renmin University, Beijing. Dr. Wang has published 45 papers and two books in water resources management. Her research interests include the river environmental flows and the relationship between water environment protection and economic development.

Xuefeng Wang is an engineer with China Water International Engineering Consulting Co. Ltd. She has previously worked as an assistant engineer within China's Ministry of Water Resources and holds a PhD in Hydrology and Water Resources from Tsinghua University.

Zhongjing Wang is a Professor of Hydrology and Water Resources in the Department of Hydraulic Engineering at Tsinghua University in Beijing. His main areas of research include regional sustainable development, water resources planning and management, hydro-informatics, macroeconomic-ecology-water resources in arid region, and water rights.

Yuan Zhang is a Professor at the Chinese Research Academy of Environmental Sciences, Beijing. Dr. Zhang has published 40 papers in water environmental management. His research interests include river environmental flows, river health and water pollution control planning and management.

Hang Zheng is a post-doctorate fellow at the Institute of Hydrology & Water Resources, Tsinghua University, Beijing. His research focus is basin water allocation planning and environmental flows. Zheng was an advisor to the China Water Entitlements and Trading project from 2006 to 2008, and was involved in the water rights pilot study in the Hangjin irrigation district in Inner Mongolia Autonomous Region.

Introduction: The Development of a Water Rights System in China

XUETAO SUN

While water shortages and the pollution of water resources are common problems around the globe, in China, the scale of these issues is immense. The demands of 1.3 billion people, together with a rapidly growing economy, have placed tremendous pressures on China's water resources. These pressures have required new approaches to the way water resources are managed. Many of these new approaches are representative of fundamental changes occurring within Chinese society, on issues such as property rights, community participation, improved environmental management and the shift towards market-based decision making.

The challenge for Chinese water managers within this setting is enormous. They must balance limited water supplies against the needs of the world's largest population; demands for rapid economic growth with calls for improved environmental management; and moves towards a market-based approach to the allocation of water with a history of state ownership and strict government control of all resources.

The 2002 Water Law of the People's Republic of China lays the foundations for improving the management of China's water. It provides for a comprehensive planning framework, addressing flood management, water resource allocation, demand planning, pollution control and other aspects of river basin management. Since the law's passage in 2002, significant steps have been taken to implement its requirements. The 'water saving society' initiative is being expanded across the country to improve water management and encourage water use efficiency. A 2006 State Council decree has formed the basis for establishing a uniform national approach to licensing and charging for water use. The master plans for China's seven major river basins are being reviewed for the first time in more than 25 years, with a view to better recognizing environmental flow requirements. A key aspect of these changes to water management has been the development of a water rights system for allocating entitlements to water and, increasingly, allowing for the transfer of water rights between different groups.

Water Rights as a Solution

In 2006, the People's Congress affirmed the importance of the development of a water rights system. The 11th Five-Year Plan of the People's Republic of China (2006–10) requires the improvement of the water abstraction license and water resource compensation system, and the establishment of a national initial water right allocation system and water right transfer system (State Council, 2006). Both the Constitution of the People's Republic of China and the 2002 Water Law provide that water resources are owned by the state on behalf of the people. However, at issue is not ownership of the resource, but the rights to allocate, to abstract and to use water. China's water rights system is its mechanism for sharing available water resources amongst competing uses and users, including providing water for the environment. This includes establishing regional rights to a share of a trans-provincial river, the rights of an individual farmer to a share of the water available in his or her irrigation district, and many other interests in between.

However, a water rights system consists of more than just the allocation process. It is designed to provide a level of certainty to water users, and other interested parties, as to how water will be managed. This in turn can allow water users to plan with some confidence as to what water they will receive in the future. Providing this certainty requires water managers to control access to and use of all natural sources of water within a basin, to regulate the operation of water infrastructure, and to understand environmental water needs. As such, water rights form the core of the water resources management system.

Development of a Water Rights System in China

The following papers consider the development of a water rights system in China. The papers review different aspects of water resources management—including water resources allocation, environmental flows, urban water management and irrigation district management—and examine how these are being addressed through a rights-based approach and how they fit within the water rights framework.

This publication is a result of a joint project between the Chinese Ministry of Water Resources and the Australian Department of the Environment, Water, Heritage and the Arts. This project, the Water Entitlements and Trading Project, has involved a review and assessment of the development of a water rights system in China using a combination of Australian and Chinese expertise. Many of the findings and recommendations of the project are documented in the following papers. The eleven papers that make up this publication broadly fall into two groups: one group covers a number of key topics from a whole-of-country perspective while the second group of articles consists of a series of case studies.

National Water Policy Issues

The first group of articles considers the different elements of the water allocation and management framework that apply across China, and how these form the basis of China's water rights system. The papers look at:

- Water resources management: This first paper provides a broad overview of the current arrangements for water management in China, including its legal and institutional framework, and policy approaches to water allocation, water resources protection and water savings.
- Water resources allocation: This paper looks at the way rights to water are shared at the basin level (amongst provinces or prefectures), at the water-abstractor

level (between different abstractors) and within irrigation districts (between farmers).

- Agricultural water management: Agriculture remains the largest user of water in China. This paper examines the different approaches taken to allocating and managing the water used for irrigation, including both ground and surface water.
- Urban water management: This paper considers the ways in which urban water needs are considered within the context of a water rights system.
- Transferring and trading water rights: This paper reviews pilot cases of the transfer of water rights—between regions and abstractors, and within irrigation districts—as a means for moving water between uses and users in fully allocated systems.
- Environmental flows: This paper looks at current methodologies applied in China for setting aside water and maintaining flows for in-stream environmental purposes. The allocation of water for environmental flow purposes is the critical first step in the allocation of water rights, as it determines the consumptive/non-consumptive balance for a basin.

Water Rights Case Studies

The second series of articles covers a number of case studies looking at the ways in which a rights-based approach has been, or could be, used to address water management challenges. The case studies cover:

- Environmental flows and water resources allocation in the Jiao River Basin: This pair of papers describes a detailed environmental flows assessment undertaken in the Jiao River Basin in Zhejiang Province. Based on the flow recommendations, the second paper demonstrates the use of scenario analysis to develop and assess options for modifying water allocations and operational rules to achieve environmental flow and water security objectives.
- Allocation of water rights in the Shiyang River Basin: The Shiyang Basin is perhaps the most water-troubled basin in China, with extreme shortages. This paper describes the development of an allocation model, for sharing water between regions in the basin.
- Allocation and management of water rights in the Hangjin Irrigation District: This paper describes a pilot study on the allocation of water rights to water user associations within a large irrigation district on the Yellow River.

The final article is a comparison between China's approach to water rights with that adopted in Australia. The paper discusses some of the lessons that each may learn from the other. Both countries have made significant progress in implementing water rights to address common (and also different) challenges.

Conclusions

Between them, these papers provide both a high-level view as well as a detailed analysis, of the different approaches being adopted in China to improve the allocation and management of water resources through a rights-based approach. No doubt, the great challenges presented by water resources management in China will continue for many years to come. However, the adoption of a comprehensive and robust water rights system should provide a solid foundation for addressing these challenges in a sustainable and equitable manner.

Reference

State Council (2006) *National Economic and Social Development Plan for Eleventh Five-Year Period, 2006–2010*, Available at: http://english.gov.cn/special/115y_fd.htm (accessed 15 November 2008).

Water Resources Management in the People's Republic of China

BIN LIU & ROBERT SPEED

ABSTRACT *Rapid economic growth in China has presented great challenges to its water resource managers due to a scarcity of water resources, severe water pollution, growing domestic and industrial water demands, and requirements for food security. This paper provides an overview of water resources in China and its management. It describes the key water issues faced by China, as well as the institutional, legal and regulatory arrangements in place to address these challenges. This includes approaches to water resources allocation and management, pollution control, and water use efficiency. The paper concludes with a discussion of the priorities and challenges for the water sector, the progress that has been made to date and the improvements that will be required to ensure the long-term sustainable use of China's water resources.*

Introduction

China is a country in a state of major transition. The country has undergone tremendous social and economic reforms over the past 30 years, and the changes are likely to continue for decades to come. The rapid economic growth in particular has presented great challenges to water resource managers, owing to growing demand for water and a scarcity of available resources, coupled with severe water pollution and other water-related environmental concerns. This paper considers these issues, and the measures being adopted to address them.

The paper starts with an overview of the state of the China's water resources, including their development and utilization. It then describes the key water issues faced by China, as well as the institutional, legal and regulatory arrangements in place to address these challenges. This includes approaches to water resources allocation and management, pollution control and water use efficiency. The paper concludes with a discussion of the priorities and challenges for the water sector, the progress that has been made to date and the improvements that will be required to ensure the long-term sustainable use of China's water resources.

While written as a paper in its own right, this article also forms an overview and introduction of sorts to a series of papers written on water resources management, particularly the development of water rights, in China. More detail on each of the topics addressed here is provided in the various issue-specific papers that follow.

Water Resources

China's 9.6 million km^2 of land mass extends from the 4000-metre-high Qinghai-Tibet Plateau in the west to the more fertile coastal plains of the east. China's scale and complex topography have resulted in a wide diversity of climates and river systems.

China's freshwater resources of 2.8 trillion m^3 amount to only one-third of the world average per capita and half the global average per hectare (MWR, 1987). These resources are unevenly distributed, both temporally and spatially. Precipitation levels gradually decline from around 1500 mm per year in the south-eastern coastal regions to below 200 mm per year in the north-western inland areas (China Factfile, 2008). Large annual variability—in the arid north, the wettest years can have eight times the precipitation of drier years—only serves to exacerbate the problems caused by the inherent shortage of water.

China's population, cultivated land and economic distribution are poorly matched with the distribution of water resources. Around 18% (134 million hectares) of the country is considered arable and permanent crop land (OECD, 2007). Two-thirds of this land lies to the north of the Yangtze River, yet this region has only 19% of China's water resources. In particular, the Yellow, Huai and Hai River Basins—which account for one-third of China's population and GDP—have only 7.7% of its water resources. Consequently, levels of water resource development differ significantly: over 90% of available water resources (including both surface water and groundwater) are utilized in the Hai River Basin and 50% in the Yellow River Basin; meanwhile less than 15% of water resources is used in the Yangtze and Pearl Basins in the south (Water Entitlements and Trading Project (WET), 2006).

China has more than 1500 rivers with drainage areas of 1000 km^2 or greater, and more than 50 000 with a catchment area over 100 km^2. The majority of these rivers are located in the eastern parts of China where the monsoonal climate produces abundant rainfall. Figure 1 shows the location of China's major river basins. Most of the large rivers have their source on the Qinghai-Tibet Plateau and follow steep trajectories down to their mouth. As a result, China has very high hydropower potential, with reserves of 680 million kilowatts (China Factfile, 2008).

Water Resources Development and Use

In the past 60 years, China has built over 85 000 reservoirs, with a total storage volume of 634.5 billion m^3. Around 75% of that total capacity comes from the 493 largest reservoirs. Across the country, supply from water infrastructure—including reservoirs, pump stations and tube-wells—is in the order of 578 billion m^3 per year (MWR, 2007). Over 80% of water use is supplied from surface water sources. However, in some northern regions groundwater contributes over 50% of total water supply (MWR, 2007).

Between 1980 and 2007, total water use increased by about 32%. During this period, domestic water demand increased by 137% and industrial demand by 236%. Secondary industries now account for around 23% of water use and domestic use absorbs around 12% (SEPA, 2006). Agriculture remains the largest user of water, accounting for nearly

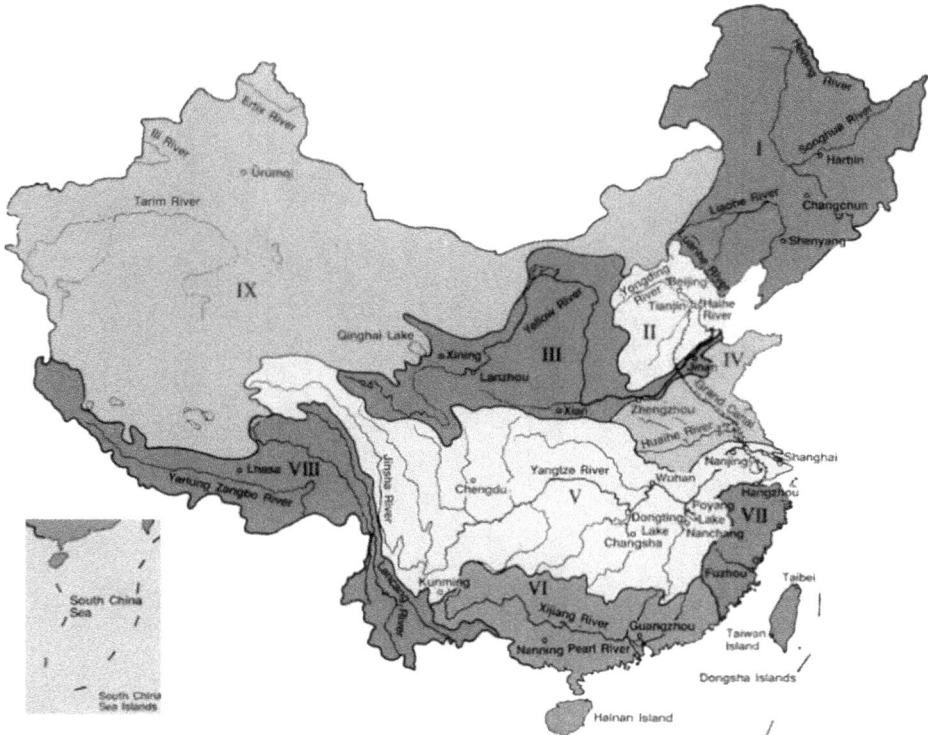

I	Songhua-Liao River Basin	VI	Pearl River Basin
II	Hai-Luan River Basin	VII	Southeast River Basin
III	Yellow River Basin	VIII	Southwest River Basin
IV	Huai River Basin	IX	Northwest River Basin
V	Yangtze River Basin		

Figure 1. Major river basins of China. *Source:* Shen (2004)

65% of all water usage. This water is used on China's 58 million hectares of irrigated land, which amounts to around 44% of China's total cultivable land. Around half of the irrigated land is located within the 5900 medium and large irrigation districts (MWR, 2007). Outside of these districts, farmers may rely on self-installed tube-wells to irrigate using groundwater (especially in the north) or supplementary irrigation from surface water supplies (Calow *et al.*, 2009). Detailed figures of total supplies and usage for each of the major river basins are shown in Table 1.

Key Water Issues

China's water-related challenges principally relate to three key areas: water scarcity, water pollution and flooding (Sun, 2002).

Table 1. Water resources and use in China's river basins

| River basin | Water resources (billion m^3) | Water usage (billion m^3) | | | | | | | |
| | | By source | | | By sector/purpose | | | | |
		Total	Surface water	Ground-water	Other type	Domestic and services	Industry	Agri-culture	Eco-environment
Total	2525.52	581.8	472.35	106.95	2.576	71.04	140.41	359.85	10.57
Songhua	92.77	40.07	23.10	16.97	0.0	3.14	7.87	28.82	0.24
Liao	38.19	20.43	8.83	11.36	0.24	3.12	2.97	14.04	0.31
Hai	24.78	38.51	12.83	25.10	0.58	5.63	5.20	26.93	0.75
Yellow	65.53	38.11	24.91	12.95	0.25	3.99	6.15	27.45	0.52
Huai	136.59	55.44	38.78	16.47	0.19	7.84	9.96	37.00	0.64
Yangtze	880.78	193.96	185.34	8.05	0.57	24.58	72.86	93.27	3.24
Southeast	179.98	33.80	32.73	0.96	0.11	4.71	11.84	16.39	0.87
Pearl	398.59	87.99	83.26	4.28	0.45	15.46	21.08	50.24	1.21
Southwest	573.91	10.87	10.54	0.31	0.02	0.99	0.77	9.08	0.03
Northwest	134.39	62.69	52.03	10.50	0.16	1.58	1.71	56.64	2.76

Source: MWR (2007).

Severe water conflicts exist between supply and demand. Based on normal demand levels and without overdrawing groundwater resources, the average annual water shortage in China is estimated to be 30–40 billion m^3. Around 400 of the 600 largest cities suffer from water shortages (OECD, 2007). At the same time, water usage continues to increase in the domestic and industrial sectors. This has resulted in water conflicts between industry and agriculture, between urban and rural areas, and between regions (WET, 2006).

Due to water scarcity and related conflicts, water is often not available to satisfy environmental requirements. Groundwater is overdrawn on average by 10 billion m^3 per year, resulting in land subsidence and seawater intrusion. As a result, in the North China Plain, groundwater tables have fallen by 10 to 50 metres over the past ten years (MWR, 2005) and parts of sections of rivers are drying up during the dry season (OECD, 2007).

Water shortages are exacerbated by high levels of water pollution. The total wastewater discharge in 2007 was 75 billion m^3, more than twice that in 1980. At the same time less than 60% of wastewater is treated. Discharge levels are making it increasingly difficult to maintain water quality and are threatening drinking water supplies. More than half of China's major cities do not fully comply with national drinking water quality standards (SEPA, 2006). Around 37% of China's land area suffers from water and soil erosion, further contributing to the degradation of river basins (WET, 2006).

As a result of pollution, around 40% of the length of China's major rivers is classified as polluted and unsuitable as a drinking water source (Table 2). Around 60% of the length of the Hai and Yellow Rivers is at this high level of pollution, with more than 50% of the Hai River so severely polluted that it is unsuitable even for agriculture.

Flood control has been a major driver to water resources management in China for centuries. However, despite huge efforts over the years in flood forecasting, planning and the construction of flood mitigation works, flooding remains a major issue on an annual basis. In 2007, classified by the Ministry of Water Resources (MWR) as a 'medium to severe' year, more than 12 million hectares was flooded. The flooding affected over 177 million people, resulted in 1230 deaths, and caused direct economic losses in the order of RMB112 billion (MWR, 2007). At the same time, water shortages have meant that water managers are now seeking to utilize flood waters as a supply source, especially in northern China.

Legal and Institutional Arrangements

Legal Framework for Water Management

The 2002 Water Law (which revised the 1988 Water Law) is China's key water legislation and provides a comprehensive framework for integrated water management (Water Law of the People's Republic of China, 2002). The law includes provisions on water resources ownership, water abstraction rights, water resource planning, water resources development and utilization, the conservation of water resources and dispute settlement. It defines river basin commissions and their responsibilities, and strengthens the administrative authority of those bodies.

The 1984 Law on the Prevention and Control of Water Pollution (subsequently amended in 1996 and in 2008) aims to prevent pollution, protect water quality and safeguard human health via, amongst other measures, regulating discharges into watercourses. The 1997 Flood Control Law establishes a framework for preventing and

Table 2. Classification of China's rivers based on water quality

River basin	Grade I	Grade II	Grade III	Grade IV	Grade V	Worse than Grade V
Total	4.1	28.2	27.2	13.5	5.3	21.7
Songhua	0.5	14.1	32.4	27.1	6.6	19.3
Liao	1.4	23.5	14.7	13.7	5.0	41.7
Hai	2.1	13.7	11.8	12.4	2.9	57.1
Yellow	3.0	13.1	27.5	15.7	6.9	33.8
Huai	0.7	12.5	24.6	18.9	9.2	34.1
Yangtze	3.2	36.2	27.5	12.4	5.9	14.8
Southeast	4.7	38.2	25.6	11.4	3.5	16.6
Pearl	0	33.1	36.3	9.7	6.2	14.7
Southwest	3.7	41.3	42.6	5.3	2.4	4.7
Northwest	31	46.5	10.2	9.0	0.4	2.9

Note: Grades I–III are suitable for drinking, grade IV for industrial and recreational use, and grade V is suitable only for agricultural use (State Council, 2002).
Source: MWR (2007).

Constitution of People's Republic of China
- State ownership of water resources

National
People's
Congress

National Laws
- Water Law
- Flood Control Law
- Law on the Prevention and Control of Water Pollution

State
Council

Administrative Regulations
E.g. 2006 Regulation on the Management of Water Abstraction Permits and the Water Resources Fee Management and Collection

Local Regulations
E.g. Anhui Province Regulation to Implement Water Law

People's Congress at the provincial level

Admin.
Department
(e.g. MWR)

Departmental Rules
E.g. Water function zones management method, 2003

Local Rules
E.g. Zhejiang Province – Detailed Implementation Rules on Water Abstraction Permit

People's Government at the provincial level

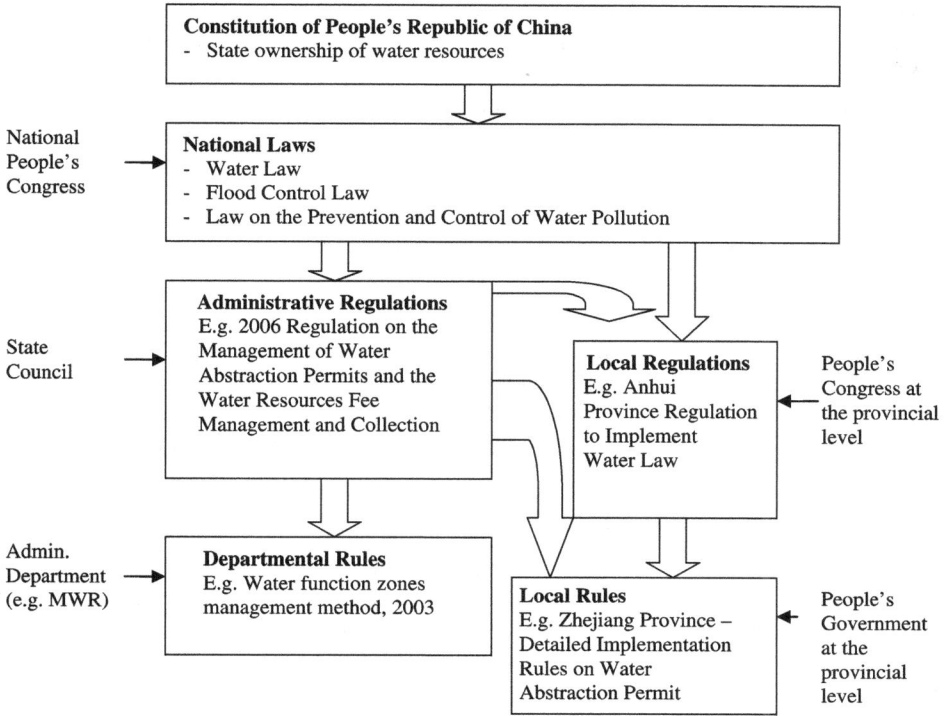

Figure 2. China's legal framework.

controlling floods, mitigating damages from flooding and water logging, and maintaining the safety of people's lives and property.

These laws were made by the National People's Congress (or its Standing Committee). Within the Chinese system, laws are supported by a system of administrative regulations (passed by the country's executive body, the State Council), local laws and regulations (passed by the People's Congress at the provincial level), and both national departmental rules and provincial government rules (made by administrative departments, such as the Ministry of Water Resources) (Figure 2). While national laws and rules set the principles, these requirements are often general in nature and local governments usually have discretion in how these are implemented (Wouters *et al.*, 2004).

Institutional Arrangements

The State Council is China's highest executive body. It comprises various administrative departments (Ministries, Commissions and other bodies), a number of which have water-related responsibilities (Shen & Liu, 2008). These include the Ministry of Environmental Protection (MEP), with responsibilities for preventing water pollution.

It is, however, the MWR that has primary responsibility for water resources management. Under the 2002 Water Law, MWR is responsible for managing water resources development, utilization, saving, conservation and flood prevention. It is responsible (at a high level) for the preparation of various plans related to different aspects of water management and for the management of water abstraction permits.

MWR consists of twelve departments (including those responsible for Planning and Programming; Policy, Law and Regulation; Water Resources Management; Construction and Management; Water and Soil Conservation; Irrigation Drainage and Water Supply; and State Flood Control and Drought Relief). Seven river basin commissions have been established underneath MWR, one for each of China's major basins. Commissions have responsibility for planning, enforcement and monitoring activities in declared major (usually trans-provincial) rivers within their basin (2002 Water Law, Article 12).

Below MWR sit provincial water resources departments, and below those—at the prefecture and county level—are water resources bureaus or, increasingly, integrated 'water affairs bureaus'. These local bureaus have functions that, at a higher administrative level, might otherwise be split across different departments. Thus, water affairs bureaus can be responsible for not only water resources management, but also water supply, urban water savings, flood control and management, and wastewater treatment (Shen & Liu, 2008).

Water Resources Management

The 2002 Water Law provides an overall planning framework, which includes a requirement for the state to prepare a 'national water plan'. This plan is given effect via a multitude of subordinate plans, prepared at both the basin and the regional level. At both these levels, 'comprehensive plans' are prepared to set the strategic objectives for water management. In addition, a number of 'special plans' are required, covering topics such as flood control, protection of water resources, development, utilization and the allocation of water. In the event of inconsistencies, a comprehensive plan takes priority over a special plan, and a basin plan prevails over a regional plan (2002 Water Law, Chapter II).

Allocation of water is via both basin and regional water allocation plans (2002 Water Law, Article 45). Through these plans, water is allocated between administrative regions. Thus, water in a trans-provincial river is allocated between provinces,[1] provincial-level plans allocate water amongst prefectures, and so on.

Water resources planning is in a developmental phase and MWR plans to establish water resources allocation plans in all major river basins. However, in some instances, such as for the Yellow River, water allocation plans have existed for 20 years, while sharing rules in the Hexi Corridor (part of the Northern Silk Road) go back more than 1000 years (WET, 2006).

Water allocation plans are given effect through a series of annual plans, which specify the available water for the year for a region or permit holder (2002 Water Law, Article 46). In some basins, annual plans are broken down to include monthly allocations, and are adjusted through the year based on climatic and flow conditions. The different aspects of the allocation framework are shown in Figure 3 (Shen & Speed, 2009).

Water abstractors require an abstraction permit to take water from a watercourse or aquifer, save some exemptions, such as for water from works built and owned by agricultural collectives (2002 Water Law, Article 7). The 2006 State Council Regulations on the Administration of Water Abstraction Licensing and Collection of Water Resources Charges ('Water Permit Regulation') sets the application and assessment process to be applied by MWR, river basin commissions, and local administrative departments in administering permits.

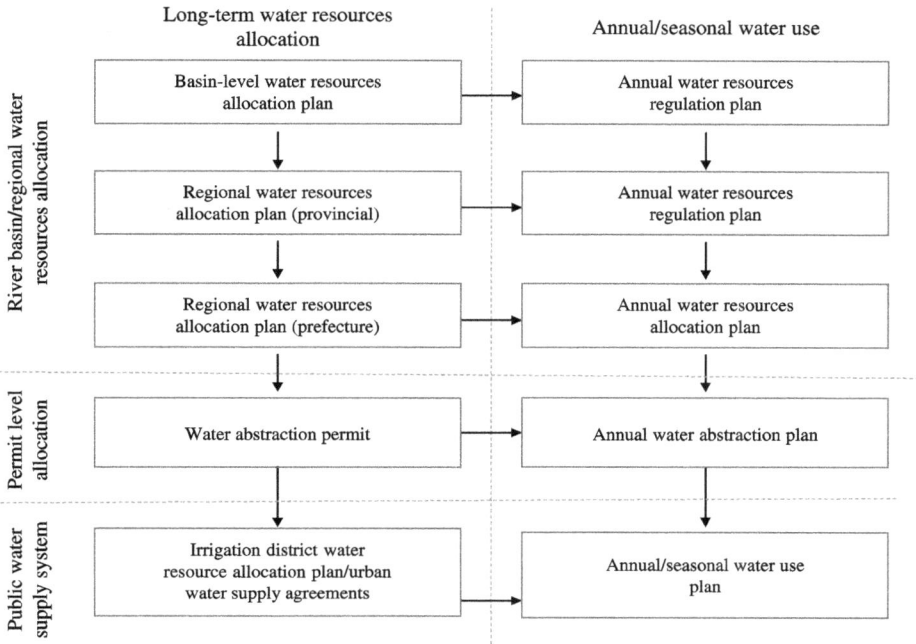

Figure 3. China's water resources allocation framework. *Source:* WET (2006)

Water abstractors must pay a water resources fee, in accordance with the Water Permit Regulation. The level of the fee varies between regions, depending, for example, on local water resources and economic conditions. Revenue from water resource fees, which was RMB6.3 billion in 2007, is hypothecated and apportioned between MWR and local water agencies (MWR, 2007).

The current framework provides some capacity for the trading of water rights. The Water Permit Regulation allows a permit holder that saves water (through changes to production processes, water saving technologies, etc.) to transfer the water "with compensation" (Article 27). The provision has been utilized in the Yellow River Basin, where lining of channels in irrigation districts has resulted in efficiency gains. This 'saved' water has been transferred to industries, which in turn have paid the cost of the channel lining (Gao, 2006). There have also been instances of local governments selling rights to water to neighbouring governments, such as in the case of Yiwu County, which purchased a permanent water right from neighbouring Dongyang County to address water shortages due to pollution (Gao, 2006).

To date, all the water transfers have either directly involved government or been driven by government initiatives. China is, however, developing systems to support a more flexible framework for trading.

Protection of water quality is managed through a zoning system. The 2002 Water Law requires MWR, via the basin commissions and local water departments, to classify rivers and lakes into "water function zones" (Article 32). Zones are set based on the purposes for which water will be used from that part of the river or lake (Figure 4), which then determines the quality of water required in the reach as well as what activities are

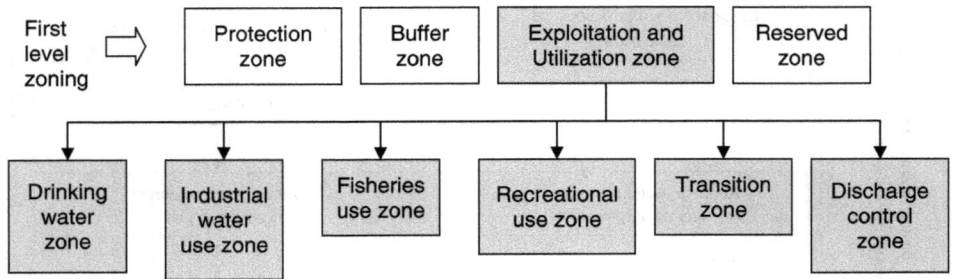

Figure 4. Categories of water function zones. *Source:* Water Function Zone Management Rules (2003).

permitted. Total waste assimilative capacity is calculated for each zone, to ensure water quality remains suitable for the designated purpose. This becomes a discharge cap for the zone. The zones are given effect through management of river discharge outlets. Applications for discharge permits are assessed to ensure that discharges are in accordance with the zone's discharge cap and other (zone-specific) limitations on particular pollutants (Water Function Zone Management Rules, 2003).

The 2002 Water Law includes a number of provisions in respect of improving water use efficiency through the promotion of water saving procedures and technologies, in both agriculture and industry (Articles 50–53). This objective is also reflected in the 11th Five-Year Plan, which promotes the dual goals of "saving resources and protecting the environment" (State Council, 2006). The concept of 'building a water-saving society' was first articulated as part of the 10th Five-Year Plan (2001–05) and the water-saving society initiative commenced in 2002 with pilot projects in three cities run under the auspices of MWR (Wang, 2003). The success of these early pilots led to an expansion of the building a water-saving society initiative. The National Plan for Building a Water-Saving Society (2005–20) sets the target of completing 10 national, 100 provincial and 1000 prefecture-level pilot programmes promoting water conservation by the year 2010. The plan also sets specific targets for the water use for different purposes.

Management Approach and Progress

China has recognized that the water-related challenges it faces demand a fundamental shift in approach to managing water resources (and the environment in general). This has involved a move away from a traditional approach of controlling or struggling against nature, to develop instead a more harmonious relationship between people and their natural environment. This has resulted in a shift away from physical solutions to these problems towards policy-based solutions. The new approach is evident in the way that water managers are increasingly seeking to:

- Address flood issues by shifting from controlling floods to managing floods (and shifting from engineering to non-engineering solutions)
- Address water shortages through the combination of 'rational exploitation' and 'building a water-saving society' (i.e. improving water use efficiency rather than seeking new water supplies)
- Rely on the natural rehabilitative capacity of nature to address environmental degradation, rather than artificial measures to restore waterways and ecosystems

• Use water function zone management to address water pollution via prevention rather than treatment.

There has been significant, quantifiable progress over past years in adopting these principles and moving towards a management regime better suited to supporting ecologically sustainable development.

Progress in Water Management

The 2002 Water Law is well recognized as providing a sound framework for the integrated management of water resources (Wouters *et al.*, 2004; WET, 2006; OECD, 2007). Between the planning systems, the water abstraction permit system, and the system of water function zones, the framework comprehensively addresses the key aspects of water resources management (Figure 5). More importantly, significant strides have been made in implementing this framework.

Over the 10 years from 1997 to 2007, the total wastewater treatment capacity in China increased from 12.9 million m^3 per day to 70 million m^3, with an urban wastewater treatment rate now of 59% (NBSC, 2008). Toxic chemical discharges from industry to rivers have been reduced by 60–70% since 1998 (SEPA, 2006). The percentage of population using improved water sources increased from 74% to 88% between 1995 and 2006. Over the same period, those with access to improved sanitation facilities rose from 53% to 65% (Millennium Development Goals Indicators, 2007).

As of 2007, 580 000 water abstraction permits were in force across the country, for a combined volume of 3545 billion m^3 (including hydropower). Water function zones have

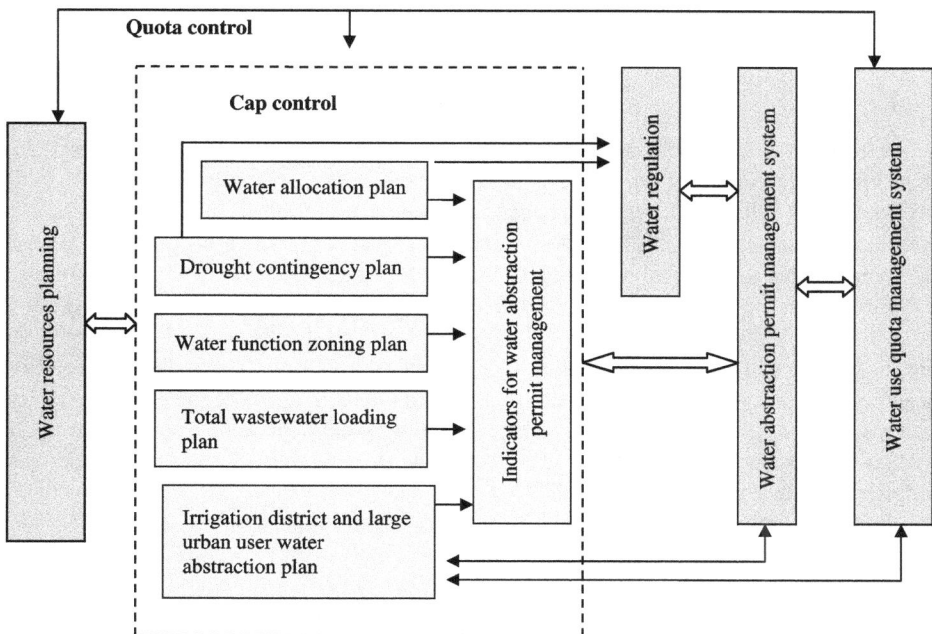

Figure 5. Water quality and quantity control mechanisms.

now been implemented in 30 provinces (MWR, 2007). Similar achievements are being made in relation to water efficiency measures. At the end of 2006, more than 100 counties had started to implement water-saving society plans, which aim to encourage reductions in pollution and more efficient water use through capacity building and innovation (MWR, 2007). Water use quotas—which set a water efficiency benchmark for different industries—have been approved for many different industry sectors, providing a scientific basis for promoting the efficient use of water resources (MWR, 2007). Community engagement in water management has also quantitatively improved. There are now more than 40 000 water user associations (WUAs) in China, with 8000 WUAs—incorporating 7 million users—created in 2007 alone (MWR, 2007). Across the country's large irrigation districts, WUAs now control nearly 25% of the irrigated area.

As a result of these efforts, while the annual GDP has grown at an annual rate of about 10% over the last 30 years, water utilization has only increased by 1% annually. The national crop output has increased by 50% while agricultural water usage has remained the same. All the while, China has continued to support 22% of the world's population with only 7% of the world's water resources and 10% of the global arable land.

Further improvements in water resources management are sought and articulated by the government in the "Decision of the State Council on Implementing a Scientific Outlook to Development and Strengthening Environmental Protection" (State Council, 2005). The decision includes a number of specific targets, to be met by 2010, in respect of: the level of urban waste treatment (increase from 59% to 70%), water consumption per unit of industrial value added (reduce by 30%), water use efficiency in agriculture (increase by 5% to 50%), the total discharge of pollutants (reduce by 10%). Administering agencies are required to develop implementation plans to achieve these objectives.

Challenges and Recommendations for the Future

Put Environmental Safety First

In the process of water development, utilization and allocation, China must pay close attention to the preservation of the aquatic environment, especially as it relates to the environmental safety and health of people. The chronic pollution of China's waterways presents its water managers with their greatest challenge. Water pollution is severely affecting people's health and quality of life. It has also become a major constraint to economic development, with an estimated loss of economic activity in the order of US$14–27 billion per year (OECD, 2007). Improved protection of drinking water supplies, particularly for groundwater, will be essential if China (and indeed the world) is to achieve its part of the Millennium Development Goals in terms of the provision of safe water supplies. China has already taken historically unprecedented steps in reducing poverty.[2] It now has the opportunity to show the world how to lift standards for safe water supplies and sanitation.

More attention should be paid to pollution protection in the western and rural areas, to prevent the polluting industries shifting from the east to west, from urban to rural areas, and generally spreading from the downstream reaches across the whole basin. The poor condition of China's waterways also requires better regulation of other water uses to minimize adverse health impacts—pollution means that both agricultural and recreational users of water can be exposed to significant health risks. Greater regulation is necessary to protect the well-being of these users. Finally, since the ecosystems of the rivers, lakes and coastal waterways in many regions are severely degraded, there is a need for greater recognition of the value of aquatic ecosystems—both directly to humans and intrinsically—and efforts need to be raised to ensure long-term ecological sustainability.

The water allocation process needs to consider the flow regime necessary to protect river health. Current approaches to defining environmental flows do not adequately account for the complex relationships that exist between the flow regime and ecosystems. Greater research is necessary to underpin environmental flow assessments, and greater awareness is necessary amongst water managers of the importance of the flow regime to river health (WET, 2007; Wang *et al.*, 2009). Management of pollution discharges will need to consider not just the water quality required to meet human needs, but also what is necessary to protect the long-term health of aquatic ecosystems.

Ensuring the Efficient and Effective Use of Available Supplies

China's water shortage problems cannot be solved through exploitation of new sources alone. Better use must be made of existing supplies. The experiences from pilot 'building a water-saving society' initiatives need to be applied widely. Importantly, greater emphasis needs to be placed on ensuring the efficient use of water, to maximize its value to society, whether economically, socially or ecologically.

Agriculture presents special challenges for water managers in China. As the sector responsible for over 60% of China's water use, and with water use efficiency in the order of 45% (MWR, 2007), irrigation must be a major focus for water management, and particularly water savings. Water-saving initiatives in Zhangye City and water transfer projects in Hangjin Irrigation District (Gao, 2006) show the potential for improving efficiency as a means to make additional water available for the environment or other sectors.

On the domestic front, encouraging more water-efficient technologies has the potential to reduce water demands and ease some of the urban water shortages. Supplying urban water needs will be all the more challenging due to continued urbanization and rising living standards. More work is required to implement the necessary standards, and also incentives to encourage a greater uptake of water-saving technologies. Ultimately, efforts in promoting efficient water use will depend on an understanding amongst water users—be they farmers, factories or the broader community—of the precious nature of water. This will require public awareness campaigns, coupled with better reporting of water facts and figures, and generally an improvement in the transparency of water policy making. The community will need to be engaged in the process if it is to become part of the solution.

Improving Water Resources Management Systems

Providing water for human and economic requirements as well as improving river health will require improvements to existing water resources management systems, both in terms of the structure of the systems themselves and their implementation. While China's laws provide high-level guidance on many of the key issues, some detailed requirements to achieve these policy goals are still lacking. Greater clarity is needed at a national level—in respect of water savings, river basin management and water resources protection—to ensure appropriate measures are adopted at a local level. Enforcement of water laws needs to strengthened, particularly to allow for settlement of inter-provincial water conflicts.

Better planning will be critical to ensuring the best use is made of available supplies. Improved water allocation and planning is required to ensure water is shared in an equitable and sustainable manner. Guidance will be necessary though, on how plans should define water allocations, how environmental flows should be specified, and generally on the legal standing of different plans. Ultimately, water allocation plans will need to become definitive tools to guide decision making, rather than aspirational targets. Importantly, plans that set economic and development objectives will need to take better account of the limitations imposed by water availability.

The rights of water users need greater clarity and security. This will provide confidence to users, encourage investment in water-dependent industries and allow for better development planning. It should also support efforts for water use to be managed within sustainable limits and reduce the potential for conflicts. While the framework for a water rights system is generally in place, much greater attention to detail in its implementation will be required to achieve the benefits that a water rights system can deliver. Improved definition of water rights will also open the way for water trading and the use of water markets and pricing mechanisms. Such systems have been successful elsewhere in promoting efficient water use and allowing water to move to higher value uses. The success of pilot projects in China suggests that water markets can also function effectively there.

Moving from Framework to Action

Tremendous progress has clearly been made in recent years. However, China's rapid development, and the sheer scale of the pollution and water shortage problems means there is still significant work to be done. There remains a wide gulf between theory and practice, and the rapid changes in management approaches have, inevitably, created issues in the policy framework that will need improvement.

Water allocation plans have not been completed in many basins, and where they do exist there can be inconsistencies between plans at the regional and basin levels. In many regions, annual regulation plans are not prepared at all, creating the potential for conflict during times of shortage (Gao, 2006). Despite requirements under the 2002 Water Law for metering and monitoring, anecdotal evidence suggests only 50–70% of water abstractors have meters installed. The largest users of water—the large irrigation district—are seldom metered. Discharges into watercourses are poorly monitored, emphasizing the challenges that remain in controlling pollution levels. The implementation of water function zones remains at an embryonic stage. While load discharge limits for the major basins were gazetted by MWR for the first time in 2008, these targets remain aspirational. Work on how to actually achieve these goals is still ongoing.

Conclusions

Protecting and restoring the environmental health of China's waterways, and thus protecting the health of its people, must be the highest priority of the country's water managers. For too long, economic development has taken precedence over environmental concerns. It is time for the pendulum to swing the other way. As development continues and the standard of living rises, the population has become more concerned with environmental conditions. At the same time, the pressure for providing high-quality water supplies and sanitation services to cities and rural regions continues to grow.

Improving water supplies and river health will require moving from a solid water management framework on paper to results on the ground. This in turn will require greater skills on the part of water managers. More sophisticated approaches to water resources management will demand a workforce that is suitably equipped to develop and implement new policies and systems. This will require a significant investment in building the capacity of water sector employees. Success will also require the understanding and support of the wider community. This in turn will require improvements in the transparency of policy making and a willingness to involve the community in the decision-making process.

The many changes that are required to improve water resources will come at a price. Improving environmental protection will require direct and prompt investment, which will have a direct economic cost. Furthermore, greater regulation is likely to have broader economic ramifications, with a loss of competitive advantage in industrial production. In improving management and regulation, it will be essential that pressures do not simply result in pollution and extraction being relocated to other parts of China. The disparity in wealth across the country makes this a real risk. The government, at a national level, needs to recognize the consequences of tighter environmental controls for undeveloped regions and provide alternatives for their future prosperity.

At the heart of the issue is, as always, the need for policymakers to strike a balance between economic development and the protection of water resources, so as to maximize the value of existing supplies while ensuring the long-term sustainability of their use. This includes determining how to balance the cost of, and responsibility for, water protection between rich and poor, between upstream and downstream, and between urban and rural areas.

Acknowledgement

This paper is the result of a project undertaken under the auspices of the Australian Department of the Environment, Water, Heritage and the Arts and the Chinese Ministry of Water Resources, with funding provided by AusAID, the Australian Agency for International Development.

Notes

1. 'Province' is used to include reference to an Autonomous Region or a centrally administered municipality.
2. There are now 200 million fewer people in China living below the poverty line compared with 30 years ago. This is a reduction of 90% in the number living below the poverty line (NBSC, 2004).

References

Calow, R. C., Howarth, S. E. & Wang, J. (2009) Irrigation development and water rights reform in China, *International Journal of Water Resources Development*, 25(2), pp. 227–248.
China Factfile (2008) *Official Chinese Government Statistics*. Available at http://english.gov.cn/2006-02/08/content_182533.htm (accessed November 2008).
Constitution of the People's Republic of China (1982). Available at: http://english.peopledaily.com.cn/constitution/constitution.html (accessed 15 November 2008).
Flood Control Law of the People's Republic of China (1997). National People's Congress of the People's Republic of China. Available at http://www.mwr.gov.cn/english1/laws.asp (accessed 15 November 2008).
Gao, E. (2006) *Water Rights System Development in China* [in Chinese] (Beijing: China Water and Hydropower Publishing).
Law of the People's Republic of China on Prevention and Control of Water Pollution (1984). National People's Congress of the People's Republic of China. Available at: http://english.mep.gov.cn/Policies_Regulations/laws/environmental_laws/200710/t20071009_109915.htm (accessed 15 November 2008).

Millennium Development Goals Indicators (2007). Available at http://unstats.un.org/unsd/mdg/Default.aspx (accessed 25 November 2008).

MWR (Ministry of Water Resources) (1987) *Water Resources Assessment in China* (Beijing: Water and Hydropower Publishing).

MWR (Ministry of Water Resources) (2005) *Report of the Ministry of Water Resources on 'Results of Investigations into National Groundwater Resources and the Environment'*, 21 April 2005 (Beijing: MWR).

MWR (Ministry of Water Resources) (2007) *Annual Water Statistics (*Beijing: MWR*)*. Available at http://www.mwr.gov.cn/english/ (accessed 20 November 2008).

National Plan for Building a Water-Saving Society (2005–2020) Issued by the National Development and Reform Commission, the Ministry of Water Resources, and the Ministry of Construction, February 9, 2007.

NBSC (National Bureau of Statistics) (2004) *China Development Report 2004* (Beijing: National Bureau of Statistics of China).

NBSC (National Bureau of Statistics) (2008) *Statistical Communiqué of the People's Republic of China for 2007* (issued 28 February 2008). Available at http://www.stats.gov.cn/english/newsandcomingevents/t20080228_402465066.htm (accessed 20 November 2008).

OECD (2007) *OECD Environmental Performance Reviews: China* (Paris: OECD Publishing).

Regulations on the Administration of Water Abstraction Licensing and Collection of Water Resources Charges (2006) State Council of the People's Republic of China. Available at: http://www.mwr.gov.cn/zcfg/xzfg/20060221000000967778.aspx. (accessed 30 November 2008).

SEPA (State Environmental Protection Administration) (2006) *State of the Environment Report* (Beijing: SEPA).

Shen, D. & Liu, B. (2008) Integrated urban and rural water affairs management reform in China: Affecting factors, *Physics and Chemistry of the Earth*, 33, pp. 364–375.

Shen, D. & Speed, R. (2009) Water resources allocation in the People's Republic of China, *International Journal of Water Resources Development*, 25(2), pp. 209–225.

State Council (2002) National Surface Water Quality Standards, BG3838 (Beijing: State Council).

State Council (2005) Implementing the Scientific Concept (Development and Enhancing Environmental Protection, State Council Document No. 39 (Beijing). Available at http://english.mep.gov.cn/Policies_Regulations/policies/Frameworkp1/200712/t20071227_115531.htm (accessed 21 November 2008).

State Council (2006) *National Economic and Social Development Plan for Eleventh Five-Year Period, 2006–2010*. Available at: http://english.gov.cn/special/115y_fd.htm (accessed 15 November 2008).

Sun, X. (2002) *Comprehensive Report of Strategy on Water Resources for China's Sustainable Development* (Beijing: The Editorial Group).

Wang, Y. (2003) The framework, approach and mechanism of water-saving society building of China [in Chinese], *China Water Resources*, 10, pp. 15–18.

Wang, X., Zhang, Y. & James, C. (2009) Approaches to providing and managing environmental flows in China, *International Journal of Water Resources Development*, 25(2), pp. 283–300.

Water Function Zone Management Rules (2003) Ministry of Water Resources. Available at: http://www.mwr.gov.cn/zcfg/20030701/65928.asp (accessed 15 November 2008).

Water Law of the People's Republic of China (2002) National People's Congress of the People's Republic of China. Available at: http://english.gov.cn/laws/200510/09/content_75313.htm (accessed 20 November 2008).

WET (2006) *Water Entitlements and Trading Project (WET Phase 1) Final Report* November 2006 [in English and Chinese] (Beijing: Ministry of Water Resources, People's Republic of China and Canberra: Department of Agriculture, Fisheries and Forestry, Australian Government). Available at: http://www.environment.gov.au/water/action/international/wet1.html.

WET (2007) *Water Entitlements and Trading Project (WET Phase 2) Final Report* December 2007 [in English and Chinese] (Beijing: Ministry of Water Resources, People's Republic of China and Canberra: Department of the Environment, Water, Heritage and the Arts, Australian Government). Available at: http://www.environment.gov.au/water/action/international/wet2.html.

Wouters, P., Hu, D., Zhang, J., Tarlock, D. & Andrews-Speed, P. (2004) The new development of water law in China, *University of Denver Water Law Review*, 7(2), pp. 243–308.

Water Resources Allocation in the People's Republic of China

DAJUN SHEN & ROBERT SPEED

ABSTRACT *Water resources allocation is a process for changing the natural or status quo distribution of water resources to meet requirements for economic and social development. The uneven distribution of water resources across China, both in space and time, makes water resources allocation of particular importance. China has developed a legal framework for the allocation of water resources that operates at three levels: at the river basin/regional level, at the abstractor level, and within public water supply systems. China has also built related systems to manage these allocations.*

Water resources allocation planning, and the implementation of related management systems, is occurring across China. However, there are significant problems in respect of how issues of integration and consistency between these three levels of allocation are addressed. Water resources allocation plans will need to be adopted as regulatory instruments, rather than the aspirational targets they currently are, to provide greater certainty for water users. Reversing the deteriorating health of rivers and freshwater ecosystems will require plans to set aside more water to meet ecological flow requirements. At a more basic level, the allocation and planning process would benefit greatly from utilization of water resources management models and increased stakeholder involvement.

Introduction

Water resources allocation is a decision-making process, involving the redistribution of water resources in respect of time (when water can be taken), location (from where it can be taken), purpose (what it can be used for), and user (who can take it). The natural spatial-temporal distribution of water resources can be regarded as the first 'allocation' of the resource; however, when this natural allocation fails to meet the needs of all water users—in terms of timing of availability, location, water quality, water quantity or reliability—there is a need to provide a mechanism for sharing the available water amongst the competing uses and users. As such, water resources allocation is a process of changing the natural or status quo distribution of water resources to balance economic and social requirements for water with other interests, including those of the environment (Shen, 2007).

Water resources planning now underpins the water allocation process—and indeed the entire water resources management systems—in many countries, including those in Europe as well as the United States, Chile, South Africa and Australia (Productivity Commission, 2003). It is the mechanism by which the sustainable limits of river basins are defined and entitlements to water are granted. Further, water resources planning provides an opportunity for an evidence-based, participatory and transparent process to allocating water resources—all of which are seen as critical to community acceptance, and hence the long-term success, of the allocation process (Hamstead *et al.*, 2008).

In China, water resources planning and allocation is increasingly being adopted as a central plank of the water management system (Wouters *et al.*, 2004). The importance of water resources plans and their role in water allocation will only increase as China moves forward in implementing a rights-based approach to water management (Water Entitlements and Trading Project (WET), 2006). Water planning is, and will remain, of particular importance in China. China is a country with enormous water resources but they are unevenly distributed in terms of their location and timing of availability. The majority of precipitation in China falls in the summer, and in monsoonal regions 60–80% of annual rainfall can come during a four-month period. Average precipitation decreases from the southeast coastal areas towards the northwest inland part of China and can range from around 2000 mm in the wet southeast to less than 50 mm in the inland desert regions (Bureau of Hydrology, 1987).

Further, water availability relative to population, arable land and economic development in China is mismatched. The area south of the Yangtze River accounts for 80.4% of the available water resources but for only 53.6% of the population, 35.2% of the arable land and 55.5% of GDP. These factors have a major bearing on the requirements for water resources allocation in China (Liu & Chen, 2001).

Legal Framework for Water Resources Allocation in China

China is, in political terms, a centralized country with five levels of administration, including central, provincial, prefectural, county and township governments, with each level answerable to the superior levels of government. China's institutional structure for water management mirrors these arrangements, with local water resources departments and bureaus answering to higher levels within the water management hierarchy, all the way up to the Ministry of Water Resources (MWR) in Beijing. MWR has overall responsibility for China's water management policy, as the water administrative department of the State Council. Under MWR, separate river basin management commissions exist for China's seven major river basins, which are responsible for management of water resources at the basin level.

Under the 2002 Water Law, a basin commission is responsible for sharing water between the provinces in which the basin is located. Likewise, within provinces, the provincial water resources department is responsible for preparing water resources allocation plans to share water amongst different prefectures, and so on down the administrative ladder (2002 Water Law, Article 45). The Water Law also establishes a water abstraction permit system to regulate the abstraction of water by individuals or entities. Further regulation of water use then occurs within public water supply systems (e.g. irrigation districts) to manage the use of water by individual users.

As such, water resources in China are allocated at three distinct, but connected, levels:

- river basin and regional water resources allocation, by which water within river basins is allocated to administrative regions and sub-regions;
- allocation amongst water abstractors, which occurs through water abstraction permits, by which a region's share of the available water resources is allocated amongst different abstractors (i.e. those who take water directly from watercourses or aquifers); and
- allocation of water amongst the end users that are supplied by public supply systems (such as those within urban water supply systems or farmers within an irrigation district).

These three stages of water allocation are discussed in detail below.

River Basin and Regional Water Resources Allocation

The 2002 Water Law provides the overall framework for water resources allocation. The macro-allocation of long-term access to water resources in China is through 'long- and medium-term water supply and demand plans'. As their names suggest, these plans broadly identify demand for water over the longer term, as well as sources to meet those requirements. They are required to be drawn up on the basis of "the current supply and demand of water, plans for national economic and social development, river basin plans and regional plans and on the principle of coordinated supply and demand of water resources, comprehensive balancing of all interests, protection of ecology, strictly practicing of economy [*sic*] and rational development of water resources" (2002 Water Law, Article 44).

Water resources are allocated in the shorter term via water resources allocation plans. These plans are prepared at the river basin level, for the purpose of regulating runoff and storage of water, and based on the relevant river comprehensive basin plan (the overarching strategic plan for the basin) and the long- and medium-term water supply and demand plan (2002 Water Law, Article 45).

For trans-provincial river basins, the water resources allocation plans, as well as drought contingency plans, are formulated by the applicable river basin management commission, together with relevant local (i.e. provincial) governments. The plans are then submitted to the State Council or its authorized department (normally MWR) for approval (2002 Water Law, Article 45). Other trans-regional water resources allocation plans and drought contingency plans are formulated by the water administrative departments of the government at the highest level necessary to have jurisdiction across the whole basin (e.g. for trans-prefecture basins, they are prepared by the provincial department), together with relevant local governments.

Water resources allocation plans are the mechanism for sharing water resources between different administrative regions within a river basin, thus identifying the share of the common resource available to those regions. Drought contingency plans typically operate in place of the water resources allocation plan during dry periods, with the trigger for their operation being a threshold defined in the plan (2002 Water Law, Article 45). Together, these plans provide a comprehensive basis for allocating water across the river basin and regions. The way water resources allocation plans define entitlements to water varies significantly across China. Some of the different approaches are discussed in more detail later in this article.

Allocation of Regional Supply to Water Abstractors

The allocation of water resources by a region to water abstractors (i.e. individuals or legal entities) is managed through a water abstraction permit system. Article 7 of the 2002 Water Law requires that the state implement a water abstraction permit system and Article 48 requires that 'units' or individuals that take water directly from rivers, lakes or underground sources must first apply for a water abstraction permit. Exemptions from this requirement exist, such as for taking a small quantity of water for domestic purposes and for watering livestock.

The State Council's 2006 Regulations on the Administration of Water Abstraction Licensing and Collection of Water Resources Charges (or the Water Permit Regulation) regulates all water abstraction facilities, including sluices, dams, canals, water pumps, water wells and hydropower stations. The Water Permit Regulation requires that the permitting system, including the grant of permits, be consistent with the various water plans required under the 2002 Water Law. This includes the water resources comprehensive plan, river basin comprehensive plan, long- and medium-term water supply and demand plan and water function zoning (which determine the suitability of water in a reach based on water quality, as well as setting restrictions on discharges to the watercourse) (Liu & Speed, 2009). Most relevantly, the grant of a permit must be in accordance with the relevant water resources allocation plan, which, at least theoretically, sets a cap on total abstractions.

In addition to considering the total volume available (as defined by the relevant plans), permit applications are considered against water use 'quotas', which are set by local water departments in conjunction with other relevant government agencies. These quotas recognize standard usage levels for certain industries or crops (e.g. for a power plant of a given size, or volume per room for a four-star hotel).

The Water Permit Regulation addresses issues including:

- responsibilities across different jurisdictions and departments for issuing permits (Article 3). Notably, different provinces set limits on which permits can be granted at different administrative levels depending on the source, the use, the total volume to be extracted, and so on (WET, 2006). In the extreme, approval may be required from the State Council, for example for projects of state significance;
- priority of supply, with domestic supply given the highest priority (Article 5). This builds on the requirements of the 2002 Water Law (Article 21), which requires that the development of water resources should first satisfy the needs of the urban and rural domestic users and "give overall consideration to the agricultural, industrial and ecological environment need for water as well as to the need of navigation";
- the application and assessment process (Chapters II and III). In addition to the Water Permit Regulation, MWR and local water departments issue detailed technical guidelines governing the application and assessment of abstraction permits. There is also a process for certifying different institutions

(e.g. consultancy bodies) as competent to undertake assessments and prepare 'water resources justification reports' for consideration by the decision-making bodies; and

- the conditions that must be included on an abstraction license (Article 24). These include volume, location, purpose, duration (between 5 and 10 years), water source and discharge conditions.

The Regulation also prohibits the granting of abstraction permits under certain circumstances, including:

- where it would increase water abstraction volumes in an area above the cap set by the water resources allocation plan;
- where significant harm might result to the water function zones (and thus the suitability of the water in that reach for its designated purpose, e.g. as a drinking water source);
- groundwater abstraction where the supply need could be met through the public water supply network; or
- where the abstraction would cause significant harm to a third party or public interests (Article 20 (2), (3), (5) and (6)).

Water Resources Allocation in Irrigation Districts and Urban Water Supply Systems

For the majority of water usage, the entities that hold the water abstraction permit supply the water they take to individual water users, rather than use it themselves. This includes public water supply companies (in urban regions) and the organizations responsible for managing and operating irrigation districts. An allocation process typically exists for assigning water to users within these systems.

In irrigation districts (often huge areas encompassing thousands or tens of thousands of individual 'farms'), water is typically allocated based on irrigation scheduling systems established during the design of the district. Generally, the scheduling system will be formulated according to reliability of supply, the irrigable land, crop patterns and crop requirements (Calow, 2009; WET, 2007). In some of the old irrigation districts, water scheduling and sharing arrangements have been in place for hundreds—even thousands—of years. In most instances, the entitlements of individuals within the system are generally well understood, although often they are not documented or protected (WET, 2006).

Urban water supply systems are generally more complicated. In addition to regulation via the water abstraction permit, a commercial contracting system is used for managing water supply and use. Typically, contracts will exist between the urban water abstractor, the water delivery company, the water treatment company and the retail company, depending on the particular organizational structure in place. The 1999 Contract Law requires that a contract exists between water users and the water utility company. Normally, this will define the service standards and charging arrangements, including the required water pressure, water quality, tariff, metering, and so on. For non-residential users, the contract will also include a maximum volume (Cosier & Shen, 2009). As part of the application process for a water abstraction permit, a public water supply organization is required to identify the purpose of use, the quantity required and the monthly use pattern. This information broadly identifies how the volume abstracted (and authorized under the permit) will be allocated within the public water supply system.

Annual Allocation of Water

The framework described above outlines the mechanism for allocating long-term allocations of water resources, typically based on average annual or monthly volumes. Water abstraction and use each year is adjusted based on actual availability via a series of annual plans.

At the river basin or regional level, an annual water resources regulation plan is prepared based on the relevant water resources allocation plans and annual inflow forecasts (2002 Water Law, Article 46). This defines the water available to an administrative region in a particular year. Similarly, the authorized water abstraction volume for a permit holder in a given year (or in some cases month) is determined according to the annual regulation plan, the forecasted inflow, water demands and the approved volume specified on the permit. Based on these elements, an annual water abstraction plan is developed to define the abstraction volume available to each permit holder.

Within irrigation districts, an irrigation water use plan is developed to manage seasonal water use. The 1999 Guidelines on Technical Management of Irrigation and Drainage Projects requires that rotational water use plans for irrigation districts be formulated yearly, while a canal water use plan and a 'grassroots' or field-level water use plan be formulated quarterly (or crop-growing period). Rotational water use plans are developed based on the soil moisture, crop type, weather conditions, and so on.

Within urban water supply systems, because of the requirements for high reliability and security, the annual water abstraction plan for an abstraction permit will normally set an annual volume that equals the volume specified in the permit (i.e. it will allow the water supply company to take its full quota). This will allow the supply company to fully meet the urban water requirements in a normal year. During drought periods special contingency rules may apply, which will provide for rationing of water (Cosier & Shen, 2009).

Figure 1 summarizes the basic water resources allocation framework in China. It shows the top-down hierarchy in place, reflecting allocations made at the level of the basin organizations (e.g. the Yellow River Conservancy Commission) down through provinces and prefectures, and ultimately to the user level. The left column reflects the long-term allocation of water rights, while the right shows the mechanism for allocating and regulating actual water on an annual or seasonal basis.

The Practical Application of Water Resources Allocation

The water allocation process is, by its very nature, an ongoing one—it requires regular adjustment to meet changed circumstances and new demands.[1] It is also time- and resource-intensive (Hamstead *et al.*, 2008). China is in the process of implementing the allocation framework described above and is well advanced in achieving this objective in a number of provinces and basins.

At the river basin and regional level, water resources allocation plans have been completed for several major river basins, including for the Yellow River, the Yongding River (for the section between Beijing and upstream provinces), the Shiyang River (sharing water between down-, mid-, and up-stream in Gansu Province), the Hei River (allocating between Gansu Province and Inner Mongolia Autonomous Region), the Tarim

Long-term water resources allocation   Annual/seasonal water use

River basin/regional water resources allocation		

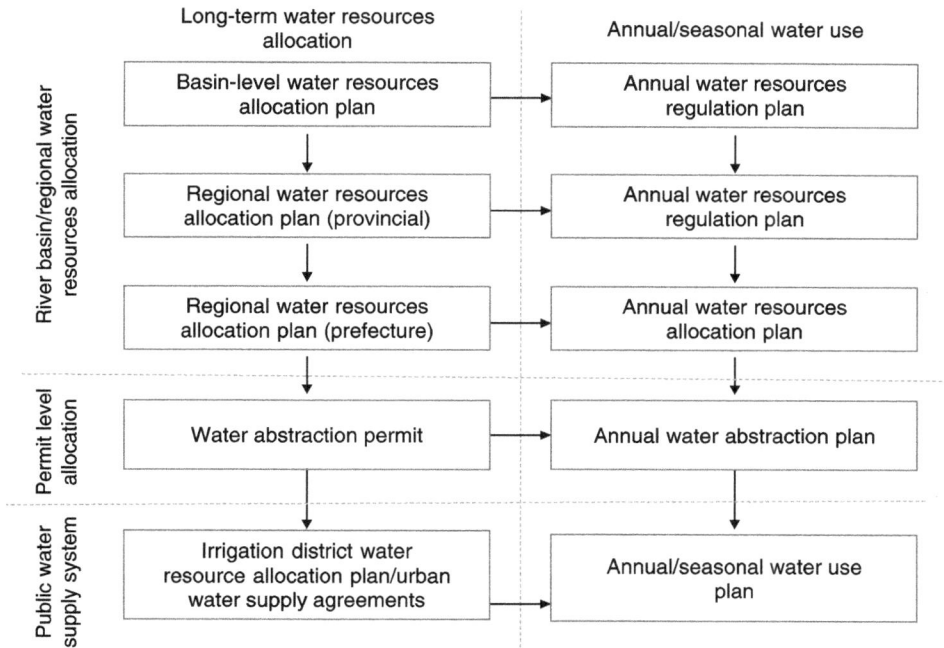

Figure 1. Framework for the allocation of water resources in China.

River (allocating between down-, mid-, and up-stream in Xinjiang Uygur Autonomous Region), the Daling River (between Hebei and Liaoning Provinces and Inner Mongolia Autonomous Region; and between prefectures in Liaoning Province), the Qiangtang River (in Zhejiang Province), the Fuhe River (in Jiangxi Province), and the Jinjiang (Jin River) (in Quanzhou Prefecture in Fujian Province). Plans have been implemented, including via water resources regulations, in the Yellow, Hei, and Tarim Rivers (Gao, 2006; WET, 2006). While the same overarching framework applies to all basins, local realities have meant that there are significant variations in respect of how these plans have been developed and applied.

Yellow River Water Resources Allocation and Regulation

The Yellow River provides the first modern example of water resources allocation and regulation in China. It continues to provide an important guide for determining how water resources are allocated and managed in river basins in China, especially in northern China.

The Yellow River is located in an arid and semi-arid zone, with an average annual runoff of 58 billion m^3 (YRCC, 2008). Since 1949, water use has increased significantly due to massive social and economic development. Starting in 1970, the downstream sections of the river experienced periods of no flow, lasting up to five years in some parts. In the early 1980s, the water diversions from the river had reached between 25 and 28 billion m^3 (YRCC, 2008), while provinces along the river required still more water to meet their social and economic requirements. At the same time, the reduced flow meant that flows were not adequate to transport the river's high sediment load to the sea.

As a consequence, and in order to resolve the resulting water conflicts, in 1987 the State Council issued the Yellow River Water Allocation Plan. The plan (summarized in Table 1) allocated a total of 37 billion m^3 of water resources amongst the ten provinces that use water from the basin.[2] The total available for allocation was determined by setting aside the 21 billion m^3 calculated as necessary for sediment transportation (YRCC, 2008). The allocation plan is based on the actual water consumption of each province. Provinces can abstract a larger volume, but are required to return much of that to the river system. (Major differences between abstraction and consumption volumes are common in irrigation districts, where large volumes flow through the channel distribution system, but are eventually returned to the river.)

Ten years after the plan was issued it had still not been implemented and continued increases in water use led to the most serious period of no flow occurring in the trunk stream in 1997. These ongoing problems resulted in the State Planning Commission and MWR issuing the Monthly Allocation of Water and Trunk Stream Regulation Plan in 1998. This defined the monthly allocation for each province in an average water year. At the same time, the agencies issued the 'Yellow River Water Resources Regulation Management Method', which detailed the process for developing the annual regulation plan. This included details on the monthly allocation principles to be applied for different water years, the locations for setting cross-boundary flow requirements, the water year (from 1 July to 30 June), and the responsibilities of different water management organizations.

Most recently, in 2006, the earlier plans were reinforced by the State Council Decree on the Yellow River Water Resources Regulation, which defined the allocation principles and procedures for the river, as well as clarified the overall responsibility of the Yellow River Conservancy Commission for preparing and implementing the plan. The decree not only reinforced earlier rules and regulations, but increased their standing due to the power of the State Council.

In accordance with these plans, decrees and regulations, an annual water resources regulation plan is prepared by the Yellow River Conservancy Commission each year, following submissions from each of the provinces. The annual water resource regulation plan works by setting:

● monthly allocations for each province (both an abstraction and consumption volume). These allocations are refined during the year by more detailed monthly plans and, during peak times, plans for 10-day periods;
● reservoir release requirements; and
● required flows at the provincial boundaries (these are the most important regulatory mechanism, and are the basis for reporting).

Due to the importance and severity of the water crisis, water resources allocation and regulation in the Yellow River has received attention at the highest political levels. The approach to water resources allocation in the basin has been an important guide for water resources management elsewhere in China and for the formulation of the 2002 Water Law. It is particularly seen as a model for water resources allocation and regulation in northern China. Several other river basin management authorities, including those for the Hei River, the Shiyang River and the Tarim River, have followed this model in terms of planning, institutional design and regulation (WET, 2006).

Table 1. Water allocation in the Yellow River

Province/region	Qinghai	Sichuan	Gansu	Ningxia	Inner Mongolia	Shaanxi	Shanxi	Henan	Shandong	Hebei & Tianjin	Total
Annual water use (billion m³)	1.41	0.4	3.04	4	5.86	3.8	4.31	5.54	7	2	37
Percentage allocated to each province	3.8	0.1	8.2	10.8	15.8	10.3	11.7	15.0	18.9	5.4	100

Source: Shen (2007).

Jinjiang Water Resource Allocation

The Jinjiang (Jin River) is located in Fujian Province in China's wet southeast. While the average annual precipitation is between 1000 and 2000 mm, water is not as plentiful as it may first seem due to the impacts of the monsoon and typhoon seasons, as well as the topography. The precipitation falls mostly between April and September—three-quarters of the year's rainfall occurs at this time—and is concentrated in the mountainous upstream areas. The combination of these factors, together with the strong economic development (which is greatest in the lowlands) has resulted in stresses on water supply, especially during the dry season.

To address the growing water conflict, Quanzhou Prefecture developed the 'Water Rights Allocation Plan for the Downstream of the Jinjiang' in 1996, which distributed water resources amongst the counties downstream of Jinji Gate (Table 2). The plan was developed based on water demands and runoff forecast for the year 2000, with 10% of the available flow reserved and 90% allocated amongst the counties (WET, 2006).

The allocation is based on runoff at 97% reliability, rather than the approach in the Yellow River of defining an annual average consumption. That is, the plan is designed to only operate in 3% of years, as it is only during the extremely dry periods that there are water shortages and some sharing mechanism is required. Additionally, the allocation is defined with reference to flow at the Jinji water gate, rather than the Yellow River approach of defining water availability for the whole river basin. As such, it is more like a drought contingency plan. This approach is common in southern China, in areas that generally have plentiful water resources but experience shortages during the dry season.

Finally, the plan is only valid for a 10-year period, which is different from plans in other basins. The plan was not triggered during its 10-year life (i.e. the 97th percentile dry year was not reached), and how well it would have worked was never tested. The plan is now being redrafted but, due to political reasons, it is unlikely that the plan will be revised; instead, regions with surplus allocation will be allowed to transfer some of their share to other regions within the basin (WET, 2006).

Table 2. Water allocation in the downstream of Jinjiang

Section	County	Allocation ratio (%)	Allocated runoff at 97% reliability (m³/s)
Southern Bank	Licheng District	2.6	1.60
	Jinjiang City	38.9	23.99
	Shishi City	10.8	6.66
	Nan'an County	6.8	4.20
	Subtotal	59.1	36.45
Northern Bank	Licheng District	12.5	7.71
	Huian County	11.2	6.91
	Nan'an City	0.9	0.55
	Xiaoxi Water Supply	5.9	3.64
	Xiutu Water Supply	0.4	0.25
	Subtotal	30.9	19.06
Reserved		10.0	6.17
Total		100	61.68

Source: Shen (2007).

Shanghai: Water Allocation within an Urban Water Supply System

Allocating and managing water supplies within an urban supply system presents different options and challenges from those that arise in allocating resources from a river basin or aquifer. Shanghai provides an example of a sophisticated approach to dealing with these issues. The Shanghai Water Affairs Bureau has overall responsibility for water resources management in Shanghai. All urban water abstractors (including both the water supply companies and individual companies that supply their own water) are required to hold, and have been issued, abstraction permits. These provide the basic restriction on the total volume of water abstracted and consumed within the city.

The Water Abstraction Office under the Water Supply Department of the Water Affairs Bureau is responsible for preparing an annual water abstraction plan. This is generally based on:

- the volumes listed in the abstraction permits;
- the actual water usage of the permit holder in the previous year; and
- the water user's proposed water usage for the coming year.

Within the urban water supply system, the water utility companies usually prepare a water supply plan, based on expected water demand as well as other factors, such as future development and the weather. Generally, an initial annual water supply plan is prepared, which is then followed up with quarterly and monthly plans.

The daily water quantity regulation of urban water supply in Shanghai is managed through a water supply dispatching and monitoring centre. The centre liaises with the bulk water supply company and waterworks treatment plants every month to determine the water regulation plan for next month, as part of the water dispatching process. At the same time, the centre regulates real-time supply according to water pressure within the supply network (Cosier & Shen, 2009).

Water Allocation and Use in Hangjin Irrigation District on the Yellow River

The Hangjin County Yellow River Southern Bank Irrigation District (HID) is famous in China's water sector as a pilot area for the transfer of water rights from agricultural to industrial users. The district has 23 000 ha of irrigable land. Under a provincial initiative started in 2003, the irrigation district saved 130 million m^3 through canal lining; this 'saved' water has subsequently been transferred to industrial users. Payments for the water rights by the industrial users have been used to fund the cost of the channel lining. Through this mechanism, the average annual take from the Yellow River by the district has been reduced from 410 million m^3 to 280 million m^3 (Gao, 2006).

To support this process, in 2005, the Hangjin County government issued the 'Water Resources Allocation Implementation Plan for Hangjin Irrigation District'. The plan allocates the water available under the district's abstraction permit to each water user association (WUA) in the district (currently there are 42 WUAs, which notionally represent the interests of farmers and are responsible for some of the local-level management). The plan sets aside water for transmission losses through leakage and for the return flow back to the Yellow River. In addition, 15% (42 million m^3) is allocated for 'ecological protection', to mitigate environmental impacts caused by the water-saving project (such as changes in groundwater and its impact on vegetation) and 10–20%

$(28-56$ million m^3) is reserved for risk management and to allow for future land development. The remainder is allocated as agricultural irrigation water rights amongst the WUAs (Zheng *et al.*, 2009; WET, 2007). The county has also established a regulation plan for dry periods, when the total diversion is below 85% of the annual average supply volume. This allows for the reduction of water available to some users, while guaranteeing supply to others.

The HID Management Bureau holds the water abstraction permit issued by the Yellow River Conservancy Commission (YRCC) and its water abstraction is therefore controlled, ultimately, by the YRCC. Annual water availability is determined through a combination of bottom-up (based on WUA requirements, calculated according to the land area to be irrigated and the types of crops that will be planted) and top-down (based on actual availability) approaches. Adjustments are made through the year to accommodate supply constraints imposed by higher-level agencies (such as YRCC). WUAs spread risk by rotational irrigation. Similarly, farmers within WUAs drawing water off a shared channel will also take rotations between years in respect of who irrigates first (with those last always running the risk of a possible adjustment by the YRCC to available supplies).

Water is allocated according to a Water Use Plan, prepared jointly by the HID and WUAs. The HID prepares a Channel Scheduling Plan of water flow at sluices leading off the main canal. Based on this volumetric schedule, Section Offices and WUAs then develop scheduling plans below this level. Following an iterative process of scrutiny and adjustment, an overall Irrigation Water Scheduling Plan for the district will be developed and submitted to higher level authorities for approval (WET, 2007).

Problems And Challenges in Water Resources Allocation

In response to water shortage issues, China has developed a water resources allocation system—centred on the requirements of the 2002 Water Law—that adopts a rights-based approach to water management, encompassing allocations at the river basin/regional level, at the abstraction level, and within public water supply systems. While on paper the framework is comprehensive, investigations have shown that in practice there are a number of issues that undermine its strength and effectiveness.

Allocations Are Not Well Specified

At present, the right being allocated is not always well defined. This is the case at all levels of allocation. At the river basin/region level, there can be discrepancies over whether allocations are for a volume that may be abstracted or consumed (i.e. with some return flow), and the duration of plans (and rights granted under them) can be uncertain. At the permit level, the abstraction right is defined simply by a volume and a basic purpose-related reliability. There are many conditions (including rules for sharing water on a seasonal basis) that are not defined, or at least not documented, and which impact on the amount available for abstraction. As such, entitlements are neither well defined nor protected from decisions that could reduce the volume or reliability of water available under the entitlement.

Within irrigation districts, and despite some efforts in pilot districts, generally allocations have not been granted at the farmer-level. The entitlements of farmers, as well as the apportionment of water for transmission losses and other ancillary requirements, are seldom specified. Recent moves to transfer water away from irrigation districts have highlighted the shortcomings of this approach (Speed, 2009).

Lack of Integration and Consistency within the System

There are inconsistencies between all levels of the allocation process. Basin and regional water resources allocation plans are typically inconsistent—in terms of either the way they allocate water or the total volume they allocate. This can be a result of the sequencing of different plans, a lack of coordination, the absence of common planning and allocation principles, or a combination of these factors (WET, 2006).

Applications for water abstraction permits are considered on a case-by-case basis because—despite the requirements of the 2002 Water Law—water allocation plans do not provide an adequate basis for making decisions relating to abstraction permits: put simply, the plans do not define rights to water in a way that enables a water manager to readily assess whether an application to abstract water will be in accordance with the grander scheme set by the overarching plan. As such, there can be inconsistencies between allocation plans (and the caps they set) and the abstraction permits granted within the plan area. In terms of annual usage, the linkages between water resources allocation plans and actual water use is weak, both at the river basin/regional level and the permit level. The processes for developing annual water resources regulation plans (based on the allocation plan) and for developing annual water abstraction plans (based on the water abstraction permits) are often not clear and in many cases these plans are not even made (Gao, 2006; WET, 2006).

Water Is Not Provided to Meet In-stream Ecological Requirements

The water allocation process is, fundamentally, the process of sharing available resources amongst different users. As such, the first stage of allocation should be to define the available resource, which in turn requires identifying the sustainable limits of a system, and hence the water to be set aside to meet in-stream ecological requirements (i.e. environmental flows). In China, the in-stream ecological water requirements are not well understood and water is generally not allocated to the environment in any meaningful way. Allocations for environmental flows (where they are made) are set as a fixed volume or a fixed percentage of the total water resources, and thus generally fail to recognize best scientific practice (Wang *et al.*, 2009).

Limited Implementation of Allocation Plans

In China, water resources allocation plans are viewed as having the status of a 'report', and are thus treated as recommendations or aspirational targets, rather than as regulatory requirements (Gao, 2006). As a result, they are often not fully implemented. There are several causes. First, the legal status of these plans is not clear or sufficiently strong. Secondly, the inconsistencies that often exist between different plans can make it difficult, if not impossible, for them to be fully implemented. And finally, both water managers and water users have often failed to place sufficient importance on these issues. It is not surprising then that there continue to be major water conflicts in China.

Improving Water Resources Allocation in China

A Regulatory Role for Water Resources Allocation Plans

Water rights systems are underpinned by water resources allocation plans (Productivity Commission, 2003; Hamstead, 2006). This is because the strength of the whole system—particularly the abstraction and use rights that sit under the plans—depends on plans allocating clear entitlements to water, coupled with adequate legal force to ensure that the plans are implemented.

In China, water resources allocation plans need to be more than just aspirational—in the sense of describing intended allocation and environmental outcomes. Rather, they should set clear requirements that can (and must) be implemented by water resource managers. These plans should provide a basis for water resource management decisions, including stating in definite terms what water can be taken, what environmental flows will be provided, and how these objectives will be achieved. The plans should set the abstraction limit for the purposes of permit management and should provide a clear basis for preparing annual water abstraction plans. This approach will provide a level of certainty to the holders of water rights at the different levels as to what water will be available for them under different circumstances. This certainty—coupled with a level of legal protection of the rights—will be important to promote confidence in the water sector, confidence that will be necessary to promote both water-dependent investment as well as any future water trading arrangements.

Incorporating Environmental Flows into Allocation Plans

Flow regimes are critical for ecological health. The requirement for environmental flows should be considered during the water resources allocation process and flows should be set based on the best available science, recognizing the importance of maintaining—to the extent possible—natural variability, in terms of duration, timing, frequency and size of flows (Bunn & Arthington, 2002). Providing environmentally relevant flows requires much more than just a fixed percentage of the mean flow.

If China wishes to ensure the sustainable use of its water resources, greater attention will need to be paid to the water requirements of its river-dependent ecosystems. In a river basin context, the allocation planning process should involve two key parts. One is the identification of key (flow-dependent) ecological assets and their flow requirements. That is, determining how the different aspects of the flow regime—such as the size, frequency and timing of flows—are important to the key ecological assets in the basin (see Gippel *et al.*, 2009a). The second part of the planning process is the identification of out-of-stream water requirements, including both volumes and reliabilities, and options for supplying this need. Water allocation planning should aim to harmonize these two key elements. An approach for balancing these two competing interests is described in detail by Gippel *et al.* (2009b).

Establishing an Integrated Approach to Allocations

Water abstraction permits should regulate the volume of water that is taken from a river or aquifer, based on limits set by a water resources allocation plan. Currently in China this is not the case. Similarly, an entitlement to take water granted under a permit—and in particular the reliability of the supply of that water—is dependent on strict controls over who else is able to take water. By linking water abstraction permits to a water resources allocation plan, the plan will be able to protect the performance (e.g. reliability) of the permit. Consequently, a strong connection between water resources allocation plans and water abstraction permits is necessary to provide for a robust water management system.

Building consistency requires two things. The first is technical consistency. That is, at a technical level, all allocations based on the same hydrological regime should be consistent in terms of the technical approach to identifying and defining entitlements. Secondly, legislation should be clear in respect of the need for consistency, and the approach to be taken to revise inconsistent plans. Water managers in China will need to work across all levels of government to develop systems capable of providing this level of consistency to be able to provide certainty and security to water users.

Application of Water Resources Management Models

Internationally, the development of allocation plans has been greatly enhanced by the use of water resources management models (Markar, 2007). These models can be used to test different water abstraction scenarios, as well as different environmental, infrastructure operational and sharing rules. This can then show the likely impacts of different allocation options on both the environment and the reliability of supply, and thus form the basis for decision-making. Currently, water planning in China is supported by hydrological—but not management—models (Markar, 2007). Through the application of whole-of-basin management models, water resources allocation plans can be developed in a way that should optimize the use of available resources and protect both environmental flows and the reliability of supply.

Promoting Stakeholder Involvement

Consultation on water allocation issues in China is typically limited to inter- and intra-governmental discussions. As such, many of the key stakeholders are excluded from the decision-making process. All parties with an interest in the allocation process should be involved in water resources planning. Such an approach should raise understanding amongst stakeholders of the water resources allocation process, as well as allow consideration of the issues of key concern to stakeholders. Experience suggests that involving stakeholders in the development of plans will help promote their implementation and ultimately improve the quality of outcomes for all concerned (Tan, 2006).

Acknowledgements

The authors would like to thank Mr. Youxing Lu, from the Quanzhou Water Resources and Hydropower Bureau, and Mr. Renliang Ruan from the Shanghai Water Affair Bureau, who supported the investigations. This paper is the result of a project undertaken under the auspices of the Australian Department of the Environment, Water, Heritage and the Arts and the Chinese Ministry of Water Resources, with funding provided by AusAID, the Australian Agency for International Development.

Notes

1. An effective water trading system may reduce the need for regular adjustments to water allocations.
2. 'Province' is used to include reference to an Autonomous Region or a centrally administered municipality.

References

Bunn, S. E. & Arthington, A. H. (2002) Basic principles and ecological consequences of altered flow regimes for aquatic biodiversity, *Environmental Management*, 30, pp. 492–507.

Bureau of Hydrology, Ministry of Water and Hydropower (1987) *Water Resources Assessment for China* (Beijing: Water Resources and Hydropower Press).

Calow, R. C., Howarth, S. E. & Wang, J. (2009) Irrigation development and water rights reform in China, *International Journal of Water Resources Development*, 25(2), pp. 227–248.

Contract Law (1999) National People's Congress of the People's Republic of China.

Cosier, M. & Shen, D. (2009) Urban water management in China, *International Journal of Water Resources Development*, 25(2), pp. 249–268.

Decree on Yellow River Water Resources Regulation, State Council of the People's Republic of China. Available at: http://www.yellowriver.gov.cn/ziliao/zcfg/fagui/200612/t20061222_8784.htm.

Gao, E. (2006) *Water Rights System Development in China* [in Chinese] (Beijing: China Water and Hydropower Publishing).

Gippel, C. J., Bond, N. R., James, C. & Wang, X. (2009a) An asset-based, holistic, environmental flows assessment approach, *International Journal of Water Resources Development*, 25(2), pp. 301–330.

Gippel, C. J., Cosier, M., Markar, S. & Liu, C. (2009b) Balancing environmental flows needs and water supply reliability, *International Journal of Water Resources Development*, 25(2), pp. 331–354.

Guidelines on Technical Management of Irrigation and Drainage Project (1999) Available at: http://www.cws.net.cn/law/guifan/SL246-1999/ (Accessed 20 September 2008).

Hamstead, M., Baldwin, C. & O'Keefe, V. (2008) *Water Allocation Planning in Australia: Current Practices and Lessons Learned*, Waterlines Occasional Paper No. 6, April 2008, Canberra: Australian National Water Commission.

Liu, B. & Speed, R. (2009) Water resources management in the People's Republic of China, *International Journal of Water Resources Development*, 25(2), pp. 193–208.

Liu, C. & Chen, Z. (2001) *The Water Resources Assessment and Supply & Demand Trend Analysis* (Beijing: China Water and Power Publisher).

Markar, S. (2007) *Evaluation of International Water Resource Management Models for Water Allocation and Planning Investigations in China*, Report for the Water Entitlements and Trading Project, Phase 2, 2007 (Canberra: Australian Department of Environment, Water, Heritage and the Arts).

Productivity Commission (2003) Water rights arrangements in Australia and overseas. Commission Research Paper (Melbourne: Productivity Commission).

Regulations on the Administration of Water Abstraction Licensing and Collection of Water Resources Charges (2006) State Council of the People's Republic of China. Available at: http://www.mwr.gov.cn/zcfg/xzfg/20060221000000967778.aspx. (accessed 30 November 2008).

Shen, D. (2007) *Theory and Application of Water Resources Allocation* (Beijing: China Water and Power Publisher).

Speed, R. (2009) Transferring and trading water rights in the People's Republic of China, *International Journal of Water Resources Development*, 25(2), pp. 269–281.

Tan, P. L. (2006) Legislating for adequate public participation in allocating water in Australia, *Water International*, 31(1), March 2006, pp. 12–22.

Wang, X., Zhang, Y. & James, C. (2009) Approaches to providing and managing environmental flows in China, *International Journal of Water Resources Development*, 25(2), pp. 283–300.

Water Law of the People's Republic of China (2002) National People's Congress of the People's Republic of China. Available at: http://english.gov.cn/laws/2005-10/09/content_75313.htm (accessed 20 November 2008).

WET (2006) *Water Entitlements and Trading Project (WET Phase 1) Final Report* November 2006 [in English and Chinese] (Beijing: Ministry of Water Resources, People's Republic of China and Canberra: Department of Agriculture, Fisheries and Forestry, Australian Government). Available at: http://www.environment.gov.au/water/action/international/wet1.html.

WET (2007) *Water Entitlements and Trading Project (WET Phase 2) Final Report* December 2007 [in English and Chinese] (Beijing: Ministry of Water Resources, People's Republic of China and Canberra: Department of the Environment, Water, Heritage and the Arts, Australian Government). Available at: http://www.environment.gov.au/water/action/international/wet2.html.

Wouters, P., Hu, D., Zhang, J., Tarlock, D. & Andrews-Speed, P. (2004) The new development of water law in China, *University of Denver Water Law Review*, 7(2), Spring 2004, pp. 243–308.

Yellow River Water Resources Regulation Management Method. Available at: http://www.hwcc.com.cn/news display/NewsDisplay.asp?id=117002 (accessed 30 November 2009).

YRCC (Yellow River Conservancy Commission) (2008) *Development and Utilization of Water Resources.* Available at http://www.yrcc.gov.cn/eng/about_yr/jj_13362425174.html (accessed 17 October 2008).

Zheng, H., Wang, Z., Liang, Y. & Calow, R.C. (2009) A water rights constitution for Hangjin Irrigation District, Inner Mongolia, China. *International Journal of Water Resources Development*, 25(2), pp. 373–387.

Irrigation Development and Water Rights Reform in China

ROGER C. CALOW, SIMON E. HOWARTH & JINXIA WANG

ABSTRACT *This article describes the growth and importance of irrigation in China in terms of the expansion of surface water irrigation led by the state, and the more recent acceleration of groundwater irrigation led by individual farmers. Key management challenges and policy priorities are outlined, highlighting the importance of water conservation and integrated water resources management under the 2002 Water Law. The article then describes the basis for rights definition and allocation planning under the Law, and recent experience with implementation in surface water and groundwater contexts.*

A key conclusion is that the development of a modern water rights system in China is vital for mediating between the claims of competing uses, particularly at the agricultural–industrial–urban interfaces, and for meeting water conservation and reallocation objectives. At the same time, farmers within irrigation districts and in emerging groundwater economies need clearly defined rights to encourage investment in the farm economy and to provide security of supply. Implementing new systems in a country the size of China is a major challenge, however, particularly across large rural aquifers where groundwater development is increasingly opportunistic and farmer-led.

Introduction

Since the reforms of the late 1970s, China's economy has grown rapidly and household incomes have risen substantially. Indeed large scale poverty reduction, particularly in rural areas, has been one of China's greatest achievements, with the number of people below the poverty line estimated to have fallen from around 250 million in 1978 to 29 million in 2003 (NBSC, 2004).[1] China's rapid growth has come at a high price to its natural resource base, however, with growing water scarcity—exacerbated by pollution—reckoned to cost China around 2.3% of GDP (World Bank, 2007a).[2] Water scarcity is especially acute in the drier north of the country, where the success of irrigation development—both surface and groundwater—has contributed to today's problems.[3] In particular, spiralling industrial and urban demands are raising difficult political questions about how to protect water-dependent rural livelihoods and meet grain targets whilst releasing water to 'higher value' municipal users.

Against this background, the 2002 publication of revisions to the 1988 Water Law marked something of a policy watershed. The revised law shifts the emphasis from supply-side solutions to integrated water resources management (IWRM) and water conservation, underpinned by a modern system of water rights. An effective system of water rights provides the basis for a number of different strategies for managing demand, including water pricing, permitting and trading. Perhaps more importantly, a water rights system provides a transparent, rules-based system for allocating water within and between uses. This is particularly important for irrigated agriculture as many of the allocation tensions now arising in China are, at their core, conflicts between irrigation and other uses, including the environment.

This paper describes how the development of a modern water rights system in China can strengthen rural people's claims to water and, at the same time, help meet water conservation and reallocation objectives. The paper begins with a brief summary of irrigation development in China, charting the rise of surface water irrigation development and the much more recent growth in groundwater exploitation. The paper then outlines some of the key management priorities that have emerged, and describes some of the more recent policy shifts that have occurred with respect to irrigation management, financing and investment priorities. The development of a modern water rights system in China is then discussed, with a review of the allocation system in general under the new Law and agricultural water rights in particular. Drawing on insights from the field, the paper then examines how policy is translated into practice, firstly in terms of the allocation of water rights to and within surface water irrigation districts (IDs), and then in terms of the allocation and management of groundwater rights. The discussion focuses mainly on experience from the drier north where competition for water, within and between sectors, is most acute. Finally, key conclusions and policy recommendations are summarized, recognizing the diversity of different irrigation 'systems' in China—from major surface water schemes to smallholder groundwater irrigation—and the need to strengthen both formal and informal rights to improve water resources management and support farm incomes.

Irrigation Development and Management Challenges

Irrigation Development

Securing food production for a growing population in the face of climatic uncertainty has long been a huge challenge for China. Although the country's land area is vast, farmland (the cultivated area) accounts for only 130 million ha, or 13.5%, of the land surface. Roughly 65% of this lies to the north of the Yangtze River, an area with only 20% of the country's total water resources but that produces around half of China's grain, including nearly all of China's wheat and maize. Irrigation development here has played a vital role in feeding the country and reducing vulnerability to uncertain rainfall. To understand how intensive development and control of water resources has arisen, however, we need to look to China's past.

The history of irrigation development in China is a long one, with records of irrigation (and flood control) stretching back over 4000 years. Successive dynasties and local rulers have organized troops and peasants to construct dykes, irrigation channels, water storage ponds and wells, first in the north of China (in the Yellow River and Huaihe (Huai) River

Basin) and then in the southern provinces to the Yangtze River. During the Tang Dynasty (818–907), for example, over 1000 separate irrigation projects were developed as state enterprises, and by the Song Dynasty (960–1297) over two million hectares of rice paddy could be irrigated under surface water schemes. Even today, irrigation and flood control works on the Min River in Sichuan Province are used much as they were originally designed (Clayre, 1984).

The Republic of China (1911–49) and the People's Republic of China (1949–present) continued Imperial China's tradition of state development and control of large works. During the 1950s, 1960s and 1970s in particular, irrigation and drainage schemes were vigorously developed. Between 1958 and 1985, for example, around RMB65 million was invested in irrigation and drainage, and between 1949 and 1996, the irrigated area increased from roughly 16 million ha to 51 million ha.[4] Over the same period, the agricultural sector experienced major institutional upheaval as rural collectivization during the 1950s—and eventually rural communization—gave way in the early 1960s to decentralization and the effective abandonment of people's communes. Subsequently, the Cultural Revolution witnessed the recentralization of farming practices and top-down controls until finally, under Deng Xiaoping, collectives were disbanded and household farming was re-introduced under the Household Responsibility System (Ash, 1993).

Initially, the break-up of collectives in the post-1979 reform period led to major increases in agricultural productivity and production as farmers regained control over land and targets were relaxed.[5] However, public investment in surface water schemes began to decline in the late 1970s as government focussed on the industrial sector and local funding for agricultural works dried up. In addition, ambiguities over system ownership and maintenance responsibilities created weak incentives for investment and upkeep which, to some extent, persist today.[6] The resulting deterioration of irrigation infrastructure, a significant decline in irrigated area in the early 1980s and declining terms of agricultural trade, contributed to the stagnation in China's grain production and a rise in food prices in the mid-1980s to mid-1990s (Lohmar *et al.*, 2003). Nonetheless, IDs of various sizes, drawing on surface water from rivers and reservoirs, still account for most agricultural water use, irrigating roughly 72% of China's irrigated land area (MWR, 2006b).[7]

The deteriorating state of surface irrigation systems in the 1980s was also a key driver of groundwater development in northern China (Lohmar *et al.*, 2003). That said, groundwater irrigation is attractive to farmers in its own right because of its reliability and 'controllability', with rural electrification providing a further catalyst for development over wide areas through motorized pumping (Calow *et al.*, 2006). The growth of groundwater-based, smallholder irrigation is therefore relatively recent, but hugely significant. In the 1950s, groundwater irrigation was virtually non-existent in northern China. In the mid-1970s, groundwater probably provided around 10–15% of irrigated supply in the water-short provinces of the north. By the mid-1990s, however, this figure had risen to around 40%, and in important downstream provinces such as Hebei, Shanxi, Henan and Shandong, where much of China's wheat is produced, the share of groundwater irrigated areas increased to around 70% (Lohmar *et al.*, 2003). As a result, Wang *et al.* (2008) suggest that, over the last 25 years, more wells have been sunk in northern China than anywhere else in the world.[8]

Up until the last 10 years or so, most groundwater development for irrigation (and domestic supply) has been publicly funded through investment in village-based, collectively owned and managed boreholes. Indeed the ubiquity of groundwater and its low development cost have made groundwater-based investment attractive to government agencies seeking quick, poverty-reducing impacts in rural areas (Box 1). More recently, however, private investment in groundwater infrastructure has increased, fuelled partly by rising farm incomes and partly in response to the falling fiscal capacity of village collectives, leading Wang *et al.* (2007a, 2007b) to conclude that private investment in and ownership of wells has become one of the most prominent features of the 'new' groundwater economy (see Box 1).[9] Today, the territory of China can be broadly divided into three types of irrigation zones, with major differences as regards irrigation dependency and the importance of surface and groundwater sources (FAO, 2004):

- The dry northwest regions and part of the middle reaches of the Yellow River, where average rainfall is less than 400 mm/year and perennial irrigation is vital for agricultural production; in these areas, major surface water diversions supply some of China's largest IDs, and groundwater development by village collectives and farm households has grown rapidly in importance over the past 30 years.
- In the North China Plain and northeast China, where precipitation ranges from 400 mm to 1000 mm/year but is monsoonal, and thus uneven; irrigation here is necessary to secure production, with 'on demand' groundwater access increasingly important in buffering rainfall variability and surface water shortages.
- In the middle and lower reaches of the Yangtze River, the Zhujiang (Zhu River) and Minjiang (Min river) and parts of southwest China, where supplementary irrigation (typically from rivers, reservoirs and ponds) is sometimes required for upland crops, and remains necessary for paddy fields, especially to increase cropping intensity.

This paper focuses principally on irrigation management in the first two zones. These are the areas where water supply is most constrained, and where there has been significantly more emphasis on improving the regulation and efficiency of water use.

Box 1. Groundwater development and cash cropping in northern Hebei. *Source:* Calow *et al.* (2006).

Across large swathes of northern China, groundwater exploitation has underpinned the intensification of agriculture, supporting farm incomes, generating rural employment and reducing poverty. Intensive development is relatively recent. In Hebei over the period 1990–2003, the contribution of groundwater to total water use rose from roughly 66% to 79%, while the contribution to irrigated agriculture rose from 63% to over 75%.

The benefits of groundwater irrigation, and the management challenges posed by dispersed rural use, are illustrated in the northern Hebei counties of Shanyi, Zhangbei and Kangbao. This area of the Mongolian Plateau, known as the Bashang, experiences long harsh winters and short humid summers, and the assurance of supply that groundwater provides for both irrigation and domestic supply is vital. Since the mid 1990s, farmers have been encouraged to shift into commercial vegetable production through subsidized well drilling and irrigation, funded by up to eight different government departments acting independently of the water administration.

While irrigated land remains a relatively minor component of the total arable area, its contribution to production and income is significant. In Zhangbei, for example, the irrigated vegetable area accounted for only 10% of the total arable area in 2003, yet contributed over 38% to total cropped production value.

Today, agriculture's share of total water withdrawals stands at around 65%, with approximately 44% of the cultivated area classified by the Ministry of Water Resources (MWR) as irrigated (MWR, 2006a, 2006b).[10] Around 50% of irrigated land falls within IDs (28 million hectares, out of a national total of around 57 million). MWR reports more than 5800 IDs with an area greater than 10 000 mu (675 ha), and 148 with an area of greater than 500 000 mu (34 000 ha) (MWR, 2006b). Most of the water used for irrigation is drawn from rivers and reservoir storage (roughly 80% in 2005),[11] although groundwater abstraction—as noted above—has grown in importance. In this respect, national averages mask significant regional and local variation in irrigation dependency and water source.

Management Challenges and Responses

The passing of China's Water Law in 1988, and its revision in 2002, signalled a renewed commitment to investment in agriculture and the introduction of new irrigation policies in the context of severe drought and growing water scarcity.[12] The 11th Five-Year Plan (2006–10) also sets out a number of policy goals and priorities for water resource management aimed at supporting rural livelihoods and encouraging the reallocation of water between sectors. These include (a) adopting a more unified management system; (b) shifting from supply-side to demand-side management; (c) integrating river basin management with regional management; and (d) establishing a preliminary system of water rights trading.

In terms of commitments, national investment in agriculture rose by 8.6% per year in the late 1980s and by 19.7% per year in the 1990s (Lohmar *et al.*, 2003). Both the 2002 Water Law and the 11th Five-Year Plan commit the government to further increases in its attempt to increase production and maintain self-sufficiency in grains,[13] raise rural incomes and reduce rural-urban inequalities. In addition, the government has taken steps to reduce the burden on farmers by abolishing a raft of locally levied agricultural taxes.[14] The political imperative to protect farmers' interests, however, has made it difficult to levy the Water Resources Fee (see below) on rural users or to achieve full irrigation cost recovery.

In terms of policy, investment priorities have shifted from new projects to the renovation and maintenance of existing surface water systems, with much more emphasis on local management, farmer participation, financing arrangements and water conservation. In particular:

- Operation and maintenance of lateral channels, water distribution and water fee collection is increasingly being taken away from village committees and put in the hands of private franchises and Water User Associations (WUAs). For example, WUAs piloted in the early 1990s in the Yangtze River Basin and Hunan Province, have received strong government backing as organs of democratic management, operating outside the traditional village-township-county line of government authority (NDRC, 2000; MWR, 2003). WUAs have been tasked with local cost recovery and channel maintenance in an effort to reduce government outlays, increase accountability between irrigation agencies, WUAs and farmers, and improve local management. The contracting of operation and maintenance responsibilities on lateral channels to private franchises—often individuals—has proved even more popular (Lohmar *et al.*, 2003; Shah *et al.*, 2004; Wang *et al.*, 2008).

- Irrigation district agencies—particularly in the north of the country—have been encouraged to operate on commercial rather than 'public provision' principles. Specifically, many agencies no longer receive core funding from government, relying instead on the fees collected by WUAs or local contractors. A common way of improving fee collection and cost recovery is to make delivery of irrigation water contingent on pre-payment (with some discretion for extremely poor households).
- Government has invested heavily in water conservation through various national schemes and programmes. For example, under the banner of 'Developing a Water Saving Society' and 'Reasonable use of surface water, limiting the use of groundwater and actively using water from the heavens', the National Water Savings Office (amongst others) in the Department of Water Resources Management (under the MWR) has implemented model projects for Water Saving Societies in a number of provinces, including the use of prepaid water tickets and intelligent card (IC) reading equipment to allocate quotas.[15]
- Embryonic water trading has been encouraged, with a number of pilot projects underway to evaluate the feasibility, costs and benefits of promoting rural-urban transfers (see Box 2). Interest in reallocation reflects, in large part, growing water shortages in China's cities. Wouters *et al.* (2004) notes that in 2004, more than 400 of China's 668 major cities were experiencing water shortages, with more than 100 cities—including major population centres such as Beijing, Tianjin, Xian, Taiyuan and Datong—experiencing 'severe' shortages.

The emphasis on water conservation and coordinated, rights-based management articulated in the 2002 Water Law also applies to groundwater management. However, while direct regulation can be pursued vigorously with those water users who are easy to identify and regulate—particularly in urban areas—dealing with large numbers of small users in rural areas is much more difficult, especially with growing private ownership of

Box 2. Changing investment priorities in Hetao Irrigation District, Inner Mongolia. *Source:* Calow *et al.* (2008).

In the Hetao area of Inner Mongolia, water has been diverted from the north bank of the Yellow River to support irrigated cropping for more than 2000 years. Indeed agriculture would be impossible in this arid area without irrigation.

Although eight large irrigation canals were constructed during the Qing Dynasty (1644–1912), the most significant investment and expansion occurred in the early years of the People's Republic. However, inadequate funding for operation and maintenance, and problems with soil salinization, led to major difficulties in the 1980s, with falls in both irrigated area and crop production.

Today, Hetao ID irrigates an area of some 8.61 million mu (0.58 million ha) with five billion m^3 of water from the Yellow River, supporting a population of over 1.2 million people. Major investment from the late 1990s to the present day of roughly RMB570 million has helped mitigate soil salinization, and the management bureau operates largely independently of government, raising its own revenues for operation and maintenance from users, organized into WUAs. Significantly, investment priorities have shifted away from system expansion to rehabilitation and water conservation, reflecting wider changes in government policy.

groundwater assets. One outcome is that groundwater development has continued apace, leading some authors to speculate that groundwater over-exploitation in northern China threatens the livelihoods of millions, and could lead to rising domestic and international food prices (e.g. Brown, 1995).[16]

What is the evidence for a groundwater crisis? A first point to note is the dearth of comprehensive, reliable and accessible data on groundwater conditions. This reflects both a lack of official monitoring, and the fragmentation of data holdings among many thousands of local and personal databases (MWR, 2001).[17] As a result, the most comprehensive assessments of groundwater conditions have been carried out through project-based field survey rather than by government agencies. One such project, discussed by Wang *et al.* (2008), involved a survey of groundwater use across six provinces, 60 counties and over 400 villages, covering the years 1995 and 2004. The authors conclude that while groundwater over-exploitation appears to be a growing problem in many villages (around 50% of the sample), problems are not universally experienced. In roughly 50% of villages, groundwater levels had shown little or no decline, and in some, groundwater levels were reported to have risen.

One area where groundwater overdraft (and pollution) clearly is a major problem is the North China Plain, home to more than 200 million people. Here, groundwater levels in the shallow aquifer have fallen by more than 15m over the past 40 years, with much greater declines in urban centres. Foster *et al.* (2004) estimate that the value of agricultural production that could be at risk from unsustainable groundwater abstraction in the depletion zone at around $840 million per year (at 2003 prices). Over such a vast area, concerted action to control pumping is required at the aquifer scale. Key questions concern the ability of government agencies to influence groundwater withdrawals, either through rights-based approaches or other mechanisms, an issue discussed further below.

Reform of Water Rights

China's decision to develop a semi-market economy and to integrate itself into the World Trade Organization (WTO) marked a break with the country's long tradition of Confucian and Socialist traditions of subordinating law to the exercise of state power (Wouters *et al.*, 2004). Since the late 1980s, China has embraced 'the rule of law' in line with western notions of the principle, creating legal frameworks in a number of resource management contexts, including water, where no frameworks existed previously (Wouters *et al.*, 2004). Hence the 2002 Water Law sets out a comprehensive framework for the planning and allocation of rights, with provisions on water resource ownership,[18] the rights of collectives to use water, water abstraction rights (both surface and groundwater), water resource planning, water resource development and use, water conservation and allocation, dispute resolution and administrative responsibilities. The allocation of water to agricultural users needs to be viewed in this broader context.

In terms of institutional arrangements, a restructured MWR, under the State Council, has the primary responsibility for water resources management, including ultimate responsibility for the preparation of water plans and the management of abstraction permits that balance demand and supply. River basin institutions and commissions,[19] such as the Yellow River Basin Conservancy Commission, are then authorized by MWR to manage all water resources (including groundwater) in their respective basins. Restructured Water Affairs Bureaus (WABs), at and above county level, are tasked with water resource administration within their political boundaries in accordance with basin plans. Importantly, WABs now manage both urban and rural water, and both surface and groundwater, under one roof.[20]

In terms of water allocation and permitting, water allocation occurs through basin and then regional allocation plans, through which water is allocated between administrative regions. Hence water in a trans-provincial river, such as the Yellow River (see below), is allocated between provinces or autonomous regions according to the overall basin plan, with provincial/regional plans then allocating water between prefectures, and prefecture-level plans allocating water between counties. In this respect, the 2002 Law builds on the basin planning model first introduced in 1998, but strengthens and extends it to all river basins (Shen & Speed, 2009).

In addition to allocating long-term rights via Water Resources Allocation Plans, a set (and related) process exists in some basins for determining the actual volume available for abstraction (or consumption) during any given year according to available and/or predicted supply.[21] The resulting Annual Regulation Plans may be adjusted during the year to bring the plan in line with actual water conditions. Hence agricultural users in an ID may be required to adjust their irrigation schedules according to available supply, particularly in drought years.

In terms of sectoral and user priorities, the 2002 Law states that the domestic water demands of urban and rural people should be satisfied first, with agricultural, industrial and environmental demands, as well as navigation requirements, considered thereafter. In practise, agricultural users in IDs have often found their entitlements curtailed first during droughts, and Lohmar et al. (2003) argues that agricultural use is typically viewed as a low priority by local government agencies keen to promote industrial development and wealth creation in 'higher value' sectors. This is also apparent in the general approach to designing new water infrastructure. Typically across China, reservoirs and supply schemes are designed to deliver urban and industrial water at a daily reliability of 97–99%, while agricultural water is usually only required to be provided at 75% reliability (WET, 2006).

Access to water resources by an individual or unit is regulated by a water abstraction permitting system, based on a regulation issued by the State Council in April 2006, and in accordance with approved water resources allocation plans and quotas. By law, all water abstractions require a water abstraction permit.[22] The 2006 regulation provides details on the process for granting and managing permits, and builds on previous (often unimplemented) permitting provisions.[23] In addition, the 2002 Law makes specific reference to the problems of groundwater overdraft, and identifies the circumstances under which groundwater withdrawals must be forbidden or restricted to ensure sustainability.

Permits for IDs are usually held by the government agency responsible for administering the district which, in turn, has its permit defined by the relevant basin commission. Farmers are then supplied a share of the water available to the ID under the permit. In some areas in northern China, an ID plan is developed specifying each WUA's or village's share of the available water. In a few pilot areas, another form of water right—referred to as water certificates—has also been granted to individual farmers, identifying their share of available water under a WUA entitlement. This is coupled with a water ticket system under which farmers pre-pay for the water they want in a particular year, season or watering. A farmer is allowed to purchase water tickets up to a limit, based on their certificate volume and seasonal availability (see below).

Holders of water abstraction permits are required to submit an Annual Water Use Plan to their administering authority. This is considered (together with actual water availability) in preparing the Annual Regulation Plan (in those basins where plans are actually made), which specifies the actual volume available to the permit holder for the year. At the level of the ID, annual or seasonal plans may then identify the actual water available that year and when it will be delivered, with some level of consultation undertaken with farmers to determine the duration and timing of irrigations. Increasingly, this consultation occurs between irrigation agencies and WUAs or, if WUAs have not yet been formed, with village committees. Again, this process is generally limited to the drier northern parts of China, with farmers granted entitlements under ID permits and annual or seasonal allocation plans, rather than formal, legally defensible rights. In many southern basins, in contrast, permits are not allocated to IDs despite the legal requirement to do so, and water use is generally based on an informal understanding and agreement of an ID's entitlement and priority (WET, 2006).

Rights Management in Practice: Insights from the Field

While the effects of the policy and legislative changes outlined above will take time to emerge, available evidence, though patchy, suggests that reforms are paying dividends. Moreover in irrigation and rights reform—as in other spheres—policymakers are learning through experimentation, showing a willingness to test alternative approaches through pilots, evaluate results and scale up what seems to work.

In this context, a growing number of projects are contributing to water rights reform and lesson learning and, although each has its particular characteristics, many share common principles. These include the assignment of water rights to specific institutions or groups within broader allocation plans; the use of water tickets and certificates to allocate volumetric rights in a transparent way; the use of consultation (in some cases) between the various water users and institutional partners; new monitoring and evaluation procedures to determine allowable withdrawals; and in some cases the use of trading to reallocate available supply between different users and uses. Specific examples are discussed below.

From River to Farm: Rights Allocation for Irrigation Districts

Rights allocation and management in the Yellow River Basin (YRB) provides perhaps the most sophisticated example of the application of a modern water rights system in China (WET, 2006). Drawing on recent work in Inner Mongolia, WET (2007) describe how rights are allocated firstly among provinces and regions according to a basin allocation plan, down to IDs through abstraction permits, and finally down to WUAs and individual farmers through informal contracts and area-based claims.

First, Inner Mongolia's share of water from the Yellow River is defined in terms of a long-term Allocation Plan, and annually through an Annual Regulation Plan. Hence in an 'average' year, Inner Mongolia receives 5.86 billion m^3 out of a total flow of 37 billion m^3. In a drought year, shares across regions are reduced, with ongoing shares within any given year detailed in the Annual Regulation Plan, published by the Yellow River Conservancy Commission (YRCC). This provides for monthly scheduling, based on monthly water use and reservoir operation plans prepared by individual provinces and, if necessary, 10-day or real-time operation.

Secondly, and within Inner Mongolia, the major surface water abstractors require water abstraction permits. Due to the large size of the entities that withdraw water there are only 17 permits, the largest of which are for the IDs of Hetao and Hangjin,[24] held by the Inner Mongolia Yellow River Irrigation Management Bureau (within the Water Resources Department). However, actual diversions under the permits are determined by an iterative process that sees bottom-up demand (from farmers and WUAs) revised through top-down adjustment and approval (under the Annual Regulation Plan). Specifically, the irrigation agency in each district prepares annual and seasonal allocation plans based on farmer consultation through WUAs, with plans then implemented through an iterative process of scrutiny and adjustment to account for supply restrictions imposed by the YRCC.

Finally, the allocation process within each ID combines bulk volumetric charging to WUAs established on branch canals with area-based charging for farmers on tertiary channels. The process sees WUAs purchase water tickets on behalf of farm members in advance of each irrigation, as part of what is both a pre-ordering and pre-payment system. Hence within each ID formal, volumetrically defined rights are not granted to individual water users because infrastructure is not in place to directly monitor flows at this level. Rather, the ID agency supplies water to WUAs on a contractual basis through annual and seasonal agreements (signed-off by agency managers and WUA chairmen, and publicly displayed), and the WUA takes responsibility for allocating shares to production teams and individual farmers. Such arrangements create a type of group right, albeit one of limited security, with individual farmers then asserting area-based claims through the WUA.[25] Farmers also have the right to cultivate land and choose what crops to plant, the right to use and repair irrigation infrastructure and the right to at least influence, through the WUA or franchise, local rules and irrigation service plans, including water scheduling.

The allocation process described above is now common in the water-scarce IDs of northern China. In particular, establishment of WUAs and private franchises has often gone hand-in-hand with the introduction of ticketing and pre-payment systems (Wang *et al.*, 2008). Field surveys in both Hangjin and Hetao IDs on the Yellow River indicate strong support from farmers, WUA leaders and agency staff, though reasons differ. For the ID agency, pre-payment systems increase cost-recovery and provide timely revenues, albeit at levels that fluctuate with ticket sales and provide little incentive to encourage conservation. Coupled with financial autonomy for the agency, there is clearly an incentive to provide a reasonable service to WUAs, and the farmers within them, to maintain revenue. Village leaders and WUAs also benefit by gaining a clear definition of their responsibilities and of the basis for financing irrigation management, and consequently a reduction in conflict at village level. For farmers, pre-payment systems also provide a direct link between irrigation charging and service delivery. Formerly, flat charges were often levied by the village leadership at the end of the year, and irrigation charges were bundled together with other payments (WET, 2007). Some commentators argue that it is new payment and incentive structures, rather than farmer participation and institutional design, that have raised the performance of IDs (Shah *et al.*, 2004; Wang *et al.*, 2008).

In some areas, efforts to improve rights definition and allocation have gone further. For example, in the case of Hangjin ID discussed above, a pilot project has recently been completed with the aim of accurately defining entitlements to water down to WUA level— the lowest volumetrically monitored points on the system (WET, 2007). In particular, a district water allocation plan has been prepared, assigning rights to water to WUAs and allocating water for distribution losses. Allocations to WUAs can now be formalized by granting water certificates to each WUA. The annual allocation process and the sale of water tickets to farmers would be undertaken within this framework.

Similar reforms have already been implemented in the water-scarce Heihe (Black River) Basin, spanning Qinghai and Gansu provinces and Inner Mongolia Autonomous Region (WET, 2006). Here, rapid socio-economic development in the river's mid-stream section led to major water shortages and disputes between upstream and downstream users, and the drying-up of important downstream ecosystems. In response, the State Council oversaw the development of a water rights system for the basin, including a water allocation plan for the main river trunk, the establishment of China's first water-saving society pilot in Zhangye, and water-saving irrigation 'sub-pilots' across several counties. In the Liyuan River ID (a typical pilot), this has involved (a) the determination of an allowable diversion for the ID, with a return flow requirement for downstream users; (b) quota allocation, within the cap, down to farmer WUAs at village level; (c) issue of water certificates to individual households, specifying rights and conditions of use; and (d) the introduction of water tickets based on annual and seasonal estimates of water availability under the permit, with provision for water trading (via tickets) coordinated through the irrigation agency.

A rather different approach to rights reform is underway in the Hai Basin Integrated Water and Environment Management Project, funded by the World Bank. A key element of this project is the distinction made between consumptive and non-consumptive water use, and the practical implications this has for the definition, allocation and monitoring of water rights when water conservation is a key objective. In particular, the distinction recognizes that only those savings in consumptive water use—specifically non-beneficial evapo-transpiration (ET) and flows to non-recoverable sinks—represent 'real' water savings which can prevent the over-exploitation of water resources within the basin. The project uses satellite remote sensing techniques to measure, target and monitor ET at the basin level and all the way down to individual farm plots, with the objective of reducing, over time, consumptive rights so defined. A similar ET-focussed approach has also been adopted on the World Bank-funded MWR Water Conservation Project to reduce the current over-exploitation of groundwater on the North China Plain (Foster & Garduno, 2004), discussed briefly below.

One objective of the Hai Basin initiative outlined above is to establish water markets based on the trading of ET quotas. Water trades of various hues have also been introduced in other parts of China, albeit on a limited basis and without a clear distinction between consumptive and non-consumptive use. In the YRB, for example, the Inner Mongolia Water Resources Department has initiated a pilot project in which downstream industries are encouraged to invest in upstream channel lining within IDs. In return, the industries receive—or bank—the water saved from reduced leakage.[26] Under the guidance of the YRCC and MWR, Inner Mongolia has now assigned water withdrawal quotas among six riverside cities, drafted a plan for water transfers, and established an Office for Water Transfer Affairs to manage transfer funds and oversee implementation. Under the scheme, Hangjin ID has transferred roughly 78 million m^3 per year to downstream industries, although rights to traded water remain ambiguous. In particular, it is unclear whether the permits assigned to IDs such as those in Hangjin are owned by the irrigation agency, or just held by such organizations on behalf of farmers in a form of 'trusteeship' (WET, 2006, 2007).

From Aquifer to Farm: The Nature and Management of Groundwater Rights

While direct regulation based on defining and managing volumetric rights can be pursued vigorously with those users who are easy to identify and regulate—such as in the case of a single off-take point for a large irrigation district—dealing with large numbers of (dispersed) small users in rural areas presents more of a challenge.[27] Hence government influence over rural groundwater use is generally much weaker than it is for IDs, where water is delivered in bulk under capped permits and then distributed to farmers (Shah *et al.*, 2004; Calow *et al.*, 2006).

What has been the experience of government agencies and village collectives with respect to rights-based groundwater management? In their review of agricultural water policy reforms, Lohmar *et al.* (2003) highlight a historical bias towards investment in and management of surface water. More recently though, they note growing interest in both the development and control of groundwater withdrawals. Through the 1980s, for example, they suggest that the monopolization of well drilling activity gave local WABs fairly comprehensive control over access to groundwater since most deep wells were sunk by drilling enterprises owned by WABs. Hence groundwater access and withdrawal rights were conferred by public agencies, with wells themselves then managed by (typically) village collectives on behalf of farmers. In more recent years, however, they note that an increase in private well drilling companies and competition among collectively owned drilling companies (by township or village) has weakened this form of control. A similar trend has also been highlighted by Calow *et al.* (2006) in the Bashang Region of northern Hebei where up to eight government departments have funded wells, often with no prior permitting approval from the relevant WAB (see Box 1). Indeed some WABs operate their own drilling companies as they are forced, or choose, to operate along commercial lines.

Reporting on progress with permitting, Foster *et al.* (2004) note that some WABs have experimented with permitting systems since the 1970s,[28] but that strong legislative support was only provided in 1993 (see above). However, most authors conclude that in rural areas, implementation has been sluggish, unrelated to resource availability, or simply non-existent. For example, Foster *et al.* (2004), commenting on groundwater development on the North China Plain, note that permitting decisions have typically been made on an ad hoc basis, with no link between groundwater availability, the development of groundwater allocation plans and permitting.

The findings of recent field surveys in northern China support the findings above, and provide additional insights into the changing nature of groundwater access and withdrawal rights. Drawing on survey results from over 400 villages and six provinces, Wang *et al.* (2008) highlight the following:

- Fewer than 10% of the well owners interviewed had obtained a permit prior to drilling, despite the near universal (local) regulations requiring one.
- Infrastructure ownership (of wells and/or pumping equipment) has shifted dramatically from collective to private as well numbers have increased. In Hebei Province, for example, collective ownership of boreholes fell from 93% in the early 1980s to 56% in the in the late 1990s. At the same time, the share of private boreholes as a proportion of the total increased from 7% to 64%.
- Informal groundwater markets have emerged in many areas, mirroring similar trends in southeast Asia. Wang *et al.* (2008) note that in the 1980s, groundwater

markets were all but non-existent; by 2004, borehole operators in 44% of villages were selling water.

In conclusion, the authors state that in most of the villages surveyed, groundwater withdrawals are 'completely unregulated' and that, in the absence of government controls, farmers have invested in groundwater infrastructure and developed informal 'spot' markets to increase groundwater access and revenue.

Such shifts are significant for several reasons. Firstly, the growth in private investment and local perceptions of access and use that are increasingly private, irrespective of the legal status of groundwater, will undermine government efforts to implement abstraction permits. Secondly, and related to this, any decline in village/collective management of groundwater infrastructure increases the complexity of management as individual interests begin to dominate (Shah *et al.*, 2004). And finally, while groundwater markets increase access to irrigation, in the absence of any volumetrically defined abstraction rights, they will tend to accelerate groundwater withdrawals (Calow *et al.*, 2006).

Against this background, what opportunities are there for introducing—and monitoring—groundwater access and withdrawal rights? Again, government agencies are responding to the problem in a number of different ways, using pilot experience to ascertain what works best in different circumstances. First, we note that government agencies can and do take drastic action to control pumping in certain circumstances. In Wuwei Municipality in Gansu Province, for example, local authorities have sealed the wells of some farmers to stop pumping altogether in an effort to restore environmental assets and stabilize groundwater levels. The greatest restrictions are in Minqin County, where excess abstractions have resulted in the groundwater level dropping at an average rate of 0.65m per year and the drying up of the terminal lakes of the Shiyang River. This county is partly irrigated from surface water from the Shiyang River, but almost 90% of agricultural water was taken from groundwater in 2006 via small tube-wells irrigating about 100 mu per well. A comprehensive package of measures has been proposed to stabilize groundwater levels. This includes a reduction in the number of permitted wells by 3000 from 9519 to 6519 (32% of the original total) over the five-year period to 2010.

This will result in a reduction of the area of irrigated farmland from 1 020 000 to 625 300 mu (a reduction of 394 700 mu, or 39% of the original area). In addition the average irrigation quota for irrigated crops over the same five-year period will be reduced from 585 m^3/mu to 476 m^3/mu (a 19% reduction). The volume pumped from the remaining wells will also be reduced so that there will be a 70% overall reduction in groundwater use, but this is partly offset by an increase in surface water allocation from 139 million m^3 per year in 2006 to 265 million m^3 per year in 2010. A greater degree of control will also be imposed on active wells through volumetric discharge regulation achieved through the use of IC cards. The reduction in irrigated areas and quotas will be accompanied by the subsidized introduction of greenhouses on a large scale which will enable a more productive use of agricultural water and increase 'crop per drop'.

In the coastal area of Zhejiang Province, around 8300 km^2 of the coastal plain has been declared prohibited or restricted for groundwater abstraction because of pumping-induced land subsidence (WET, 2006; Wang *et al.*, 2009). And in some areas of the Bashang Plateau in northern Hebei (see Box 1), well drilling in counties facing severe groundwater drawdown has been prohibited under new regulations (WAMH, 2007). In view of the government's increasing concern with farmers' livelihoods and alleviating the peasants' burden, however, such moves may be exceptional.

Secondly, major project-based investment in registration and permitting has occurred in some over-exploited areas, supported by donor agencies and government partners. For example, the World Bank-funded Water Conservation Project on the North China Plain has invested heavily in county-level capacity building (for WABs) and farmer/field level education and technical change, focusing particularly on (a) technical, management and agronomic measures for reducing ET and generating 'real' savings; (b) groundwater monitoring; and (c) the monitoring and enforcement of abstraction quotas (Foster & Garduno, 2004). A recent evaluation (World Bank, 2007b) indicated that these measures had significantly raised agricultural production and farm incomes for over 300 000 households, while reducing groundwater overdraft and non-beneficial water losses.

Also in Hebei, the provincial WRD has established Integrated Water Saving Demonstration Projects in eight counties, combining similar water-saving measures (technical, agronomic, water scheduling/management) with groundwater management WUAs. Using IC card-reading technology—growing in popularity in China—farm households are assigned water quotas and water certificates (based on land holdings and crop water requirements) and can only irrigate with prepaid cards. WUAs then assume responsibility for the upkeep of completed systems, collect maintenance fees, organize rotations, monitor quotas and provide incentives and penalties for below and above quota abstraction, respectively (Calow *et al.*, 2006). In this way, government agencies set the rules of the game and provide technical support, but leave detailed monitoring and enforcement of household rights to farmers themselves. Whether such approaches can be scaled up to affect aquifer-wide groundwater conditions is debatable, however. In particular, the co-existence of village-based WUAs operating within a well quota and private entrepreneurs operating outside it opens the door to 'free-riding', with the conservation gains of the former simply captured by the later.

The Water Resources Demand Management Assistance Project (WRDMAP), funded by the UK Department for International Development (DFID), has advised on measures to improve groundwater management in Wuwei Municipality in Gansu Province. The large reduction in groundwater use needed to stabilize groundwater level conditions and thereby address to some extent environmental concerns was noted earlier. This will be achieved by revising well permits, issuing household water rights certificates (based on reduced norms for crop water requirements), installing IC card technology and increasing water fees. These restrictive measures will be accompanied by improvements to water delivery infrastructure, a smaller increase in surface water allocation, assistance with new crops and agricultural/irrigation techniques (including subsidies for greenhouses and drip irrigation). Institutional arrangements are also being strengthened through establishment of a management bureau at river basin level and WUAs at village level. Each village (or WUA) typically includes 20–50 boreholes which are managed by production groups (water user groups). The tasks of the village-level WUA include assistance to the water management station (WMS) in many of the new groundwater management responsibilities, including the issue of household water rights certificates, enforcement of permits and the collection of fees. These are onerous requirements and thus the WUAs are repaid part of the water resource fees collected in order to cover a small salary for directors and vice-directors and some administrative costs—in recognition of the role that WUAs play in water resources management. This formal process of paying staff from part of the newly-introduced water resources fee is important for ensuring that WUAs are effective and sustainable.

Household water rights certificates were introduced in 2007. These are prepared by the water management station and WUA and make allowance for all uses of water (domestic, agricultural and livestock). The precise sharing of responsibilities varies between counties and is being revised as experience is gained—the intention is generally that the WUA should take the primary role for entering correct household data, calculating the water rights on the basis of norms issued by the county WAB, and issuing the certificate on behalf of the WAB. The WMS provides a monitoring and quality assurance role as well as technical assistance. The certificates also allow for recording actual water use and thus serve a dual role of ensuring that the farmer receives water in accordance with their right and that abstractions are controlled in accordance with the norms. The certificate, however, does not yet mention the source of water, as the right is intended to be independent of the source. This causes some complexity in monitoring—particularly in areas which can receive both surface and groundwater. The certificate is a household certificate, held by the head of the household. This is significant as there is considerable off-farm employment and migration, and the (usually male) heads of household are often away.

IC cards were introduced on a pilot scale from 2006, and were widely installed in 2008. The intention is that all agricultural boreholes in Wuwei municipality should be fitted with IC card systems by 2010. These are installed by the WMS but maintenance is the responsibility of the WUA. In this case, the IC card is for the borehole rather than the household and limits the total abstraction to the sum of the household rights relating to that well. It should be noted that households often have land in more than one borehole command area, and this has to be taken account of by the WMS when charging the IC cards. Internal distribution within each borehole command area continues to be managed by the well operator or production group leader. This is a relatively simple task, but it will become increasingly contentious as norms are reduced and the WUA will need to be active in monitoring compliance with rights and resolving any conflicts arising from failure to observe them. The alternative which some farmers would prefer—individual IC cards corresponding directly to their water right, which they could use on any well—is considered too complex to administer in the short term, but may be introduced later.

The combination of rights certificates and IC cards has been introduced rapidly, and it is not surprising that there have been teething problems and a high workload for water management staff. The detailed working procedures are still being developed and modified in response to difficulties encountered. Well permits are now being re-issued by the Shiyang River Basin Management Bureau following a survey of all wells which verified location, depth, pump capacity, pump/well age and condition, area irrigated and other uses of water. This data has been entered into a geographical information system (GIS), together with some socio-economic and other data. Permits are currently issued for each well, although options for issuing a single permit for a WUA, to cover a group of wells, are being considered for the future.

Water resources fees for groundwater were introduced in 2007; these are intended to recover the costs of managing the resource (issuing and monitoring permits etc). Fees for excess water withdrawals are charged at a higher rate, but the fee is so small compared to the electricity charge that it does not act as a disincentive: control of water use thus relies on strict administrative enforcement of the water rights (via the use of IC cards and ensuring that pump operators comply with irrigation schedules agreed with the WMS). Fees are paid in advance and 'water tickets' are then issued. In theory these can then be traded (for up to three times the face value of the ticket) but this has not yet happened. In Minqin County, 5% of the water resources fee is returned to the WUA to cover part of their costs. Although water fees are prepaid, the electricity charges (which are the main cost incurred for groundwater irrigation) are paid in arrears and thus the burden on farmers is not excessive.

Finally, in the Bashang region of northern Hebei, we note that similar pilots have built on groundwater resource assessment, use/user surveys and risk mapping to determine which areas require high priority management of the kind described above. In addition to the targeting of WUA-based pilots, the assessment and mapping approach has also prompted Zhangjiakou City government to issue a number of local regulations intended to provide 'teeth' to the broad provisions of the 2002 Water Law. Hence, new regulations make it clear that county and city WABs need to approve any new well drilling across government departments in accordance with the groundwater maps described above. Perhaps more significantly, wider shifts in economic policy are also occurring, with an emphasis on livestock rather than (irrigated) cash crops. In view of the limited coverage of WUA-IC pilots, the growth in private investment in groundwater and the challenges of direct regulation, these changes may be more effective in relieving pressure on the resource base than narrow water resources management (Calow *et al.*, 2006).

Conclusions and Recommendations

In order to meet its growing demand for water with limited water resources, China needs to modernize and reform its system of water rights and establish effective implementation of such a system. A water rights system provides a transparent, rules-based system for allocating water within and between uses. This is particularly important at the sectoral level because irrigated agriculture is under growing pressure to release water to urban and industrial users. At the same time, farmers within IDs and in emerging groundwater economies need clearly defined rights so that they can make long term investments in agriculture, secure in the knowledge that their rights can be defended against competing claims except in exceptional circumstances.

This article has described the growth and importance of irrigation development in China, firstly in terms of the expansion of surface water irrigation led by the state, and secondly in terms of the more recent acceleration of groundwater irrigation, led increasingly by individual farmers. Key management challenges and changing policy priorities have been outlined, highlighting the growing emphasis on water conservation and IWRM under the new Water Law and the 11th Five-Year Plan. The article then examined the basis for rights definition and allocation planning under the Law, looking in detail at how formal rights are allocated to agricultural users under basin allocation plans in both surface and groundwater contexts. Finally, the article has described implementation experience, firstly in terms of rights allocation to and within IDs, and secondly in terms of groundwater access and withdrawal rights.

Before looking at specific recommendations, some general observations are made. First, it is clear that China's quarter-century record of economic dynamism has been built on a willingness to experiment with new reforms in a pragmatic and flexible way. This is also apparent with water reform, as reference to China's many pilot projects in this paper has highlighted. Secondly, it is important to recognize a diversity of different water rights when discussing water resources management, looking beyond formal, state-issued authorizations to informal entitlements brokered by local organizations. At a local level, rights are embedded in contracting arrangements within IDs, in the collective property rights of village-managed groundwater systems, and in the allocation of household abstraction quotas under emerging groundwater-based WUAs. They may work well in managing local claims and obligations, especially where formal rights cannot easily be monitored and enforced. Finally, while this article has drawn heavily on experience from northern China, the challenges of water resources management in the wetter south should not be under-estimated. Here too, growing demand for water raises similar (albeit less pressing) questions about rights allocation, and the recommendations outlined below should be viewed in a national rather than regional context.

Drawing on the review of Chinese experience in this chapter, but also on wider international experience in rights reform, the following recommendations are offered.

Ensure an integrated and consistent approach to rights definition and allocation. Arguably the most detailed and robust systems have developed in those regions where water is most scarce, particularly in the north of China. The process of rights allocation within the Yellow River described in this paper demonstrates how integrated planning between different levels, sectors and users can occur, such that the water entitlement of a farmer within an ID is linked to the rights held by the district as a whole, and to each region's share of the available water from the Yellow River. Similar systems should be introduced in those basins where rights-based management has not yet been developed or implemented, including those in the south of the country.

Adopt water resource allocation plans as the basis for defining and allocating water rights, including the permits of IDs. The Yellow River case study also demonstrates how allocation plans should be developed, at basin and regional levels, as a basis for allocating water within a basin. Hence, in the Yellow River Basin, the plans clearly identify the water available for abstraction by IDs and other major users under variable resource conditions, and set capped permits that are strictly monitored and enforced. Similarly, allocation plans should specify groundwater allocations based on estimates of clearly understood aquifer safe yield as a basis for permitting and other management efforts (see below). In many basins, however, information on groundwater conditions remains limited, there is little or no connection between allocation planning and permitting, and permits (if issued at all) are often granted on an ad hoc basis.

Ensure that allocation plans adopt consistent terminology and planning processes with respect to consumptive and non-consumptive uses. In the south of China, the permits issued to IDs often prescribe an abstraction volume. In the drier north, a consumption volume with end-of-system return flows may be specified. Where practical, a distinction should be made between consumptive and non-consumptive use so that return flow accounting informs the overall allocation plan and issue of downstream rights. In this respect, water conservation and transfer programmes targeted at the agricultural sector require careful design. In an ID where irrigation returns provide useful aquifer recharge, for example, allowing irrigators (or an agency) to sell or lease rights defined as diversions may severely affect the rights of downstream users (see below).

Explore opportunities for providing certainty and security for holders of agricultural water rights granted by the state. The process for granting water rights, and in particular for allocating water to IDs on an annual basis, should be clear and consistent so that ID agencies—and the farmers within them—know when water will be available, how much water they will get and for how long they will get it.

Strengthen the claims of farmers within IDs through user-group contracting, the allocation of water certificates and increased involvement in local decision-making. Within IDs, WUAs and contractors also have an important role to play in ensuring that farmers receive timely information and can enforce their entitlements. Management reforms that increase participation in and the accountability of WUAs, and that support transparent and incentivized contracting arrangements, therefore play an important role in strengthening farmers' water rights. WUAs have been introduced rapidly over much of the country, particularly the north, but not all have been as effective as the initial pilot associations. Further measures to support and strengthen these WUAs, particularly for groundwater management, are needed if they are to become sustainable local organizations.

Explore opportunities for water trading but recognize its limitations. The sale or lease of agricultural water rights can raise the overall productivity of and returns to water, and can generate significant economic benefits once certain preconditions are in place. Establishing clear, enforceable rights, and then developing effective markets that work in the public interest, is a major challenge. Clear criteria for approving transfers and for predicting (and addressing) third party impacts are required for larger-scale, regional and/or inter-sectoral transfers. The growth of groundwater markets has been farmer led, with no volumetric caps on total withdrawals. Such markets can increase access to water and expand irrigation, but are likely to increase overall abstraction and accelerate overdraft in vulnerable areas.

Recognize the importance of groundwater irrigation in water resources planning and management. Groundwater development has played a significant role in increasing agricultural production, raising farm incomes and supporting rural development, particularly in the north of China. Yet its contribution to livelihood support, and the need for proactive management, has gone largely unrecognized by a water bureaucracy that often focuses on surface water engineering. The result, in many areas, has been explosive and sometimes unsustainable development, patterns of groundwater abstraction that are very difficult to regulate (particularly where private interests are entrenched), and allocation licensing that, if present, is unrelated to resource availability.

The starting point for considering groundwater management options is an understanding of patterns of use, and of services that need to be protected and that are feasible. 'Thick and deep' approaches to groundwater management, based on well permitting and volumetric licensing, may be difficult to apply across the large aquifers at risk in many rural areas of northern China. There are exceptions, of course, but these have typically involved either major project investment (difficult to scale up) or a reaction to emblematic events (isolated and infrequent). In this context, management approaches based on intensive, rights-based regulation by government agencies may be best directed at protecting 'strategic' aquifers where domestic use is threatened, or where reserve supplies may be needed to protect against shortages in extreme drought conditions. These may include deeper, confined aquifers that store potable water and have not yet been tapped by irrigators. In other areas, 'thinner' approaches may be more appropriate: support for groundwater management WUAs that allocate household quotas through IC equipment, for example, or the licensing of drilling companies rather than abstractors. Where private rights are now firmly entrenched, however, conventional approaches predicated on hydraulic control and regulation may offer little leverage. In these circumstances, government effort might be better directed at influencing wider economic incentives, in particular, providing incentives for less water-intensive cropping or, conceivably, supporting shifts into the rural non-farm economy.

Acknowledgements

This paper is the result of a project undertaken under the auspices of the Australian Department of the Environment, Water, Heritage and the Arts and the Chinese Ministry of Water Resources, with funding provided by AusAID, the Australian Agency for International Development.

Notes

1. Using China's standard of defining poverty, i.e. income below $0.2 per person per day at the current exchange rate, or less than $0.6–0.7 at purchasing power parity (PPP). Note that China's poverty line is lower than that used by the World Bank to measure poverty in other countries ($1 per day at PPP).
2. The total environmental damage costs of air and water pollution were estimated at 5.8% of GDP (World Bank, 2007a).
3. Northern China is generally referred to as the area north of the Yangtze River. In terms of water availability, the North China Plain (or 3H Basin) has only about one-third the national average and about half the per capita water availability specified by the UN as the standard for maintaining socio-economic and environmental development.
4. Spence (1999) reports that during a three-month period between late 1957 and the end of January 1958, the State mobilized 100 million peasants to create a functioning irrigation system for 7.8 million hectares of land. Bramall (2000) reports that the proportion of irrigated land rose from 20% in 1952 to 50% in 1978.
5. Mandatory targets governed sown areas, yields, levels of input applications, planting techniques and other factors on a crop-by-crop basis. After 1978, mandatory targets were replaced by 'guidance planning' and market allocation (Ash, 1993).
6. De-collectivization and the re-introduction of household farming led to uncertainty about who should own and manage irrigation infrastructure and contribute to maintenance, especially as local government was reluctant to take charge of (and therefore subsidize) irrigation projects. Water charging—at a very low level—was only initiated in 1980. Hitherto, only a few larger and older IDs charged nominal fees; in most schemes, farmers paid no charges but were expected to contribute labour for construction and maintenance (Stone, 1993).
7. MWR (2006b) indicates that medium-sized IDs of over 10 000 mu (667 ha) and large IDs of over 300 000 mu (20 000 ha) account for 72% of the effective irrigated area.
8. According to official estimates, the number of wells in all of China was roughly 150 000 in 1965. By the late 1970s there were more than 2.3 million, and by 2003 the number had risen to around 4.7 million (reported in Wang *et al.*, 2007b).
9. Government abolition of local taxes and levies on farmers by village and township leaders may also undermine collective investment in groundwater infrastructure and indirectly encourage private development.
10. In practice the irrigated/rain-fed distinction is uncertain because of changing climatic conditions and irrigation needs from different sources.
11. MWR, 2006b. In reality the surface water/groundwater partition is also uncertain because of (a) the significance of informal, unmonitored groundwater development; and (b) the fact that groundwater demand will vary with surface water availability and climatic conditions, particularly in those areas that rely on groundwater as a supplemental or buffer source.
12. Revisions were required to address the growing problems of water scarcity and pollution that had arisen in the 1980s and 1990s. The 2002 Law makes water resource conservation a general principle in all relevant areas, and the subject of 17 separate articles (Wouters *et al.*, 2004).
13. The Chinese government maintains an unwritten policy of ensuring roughly 95% self-sufficiency in grains to ensure an adequate supply of affordable food. Agricultural trade broadly reflects comparative advantage, with sharp rises in imports of land intensive oil crops rather than wheat, rice or maize (OECD, 2005).

14. Since 2000, the government has attempted to reduce the 'peasant burden' by phasing out a range of government taxes, township and village levies and miscellaneous fees. In 2004, the government announced the phasing out of the Agricultural Tax over a period of five years (OECD, 2005).
15. According to the Ministry of Water Resources (MWR, 2006a), the 'water saving' irrigated area now comprises 34.5% of the total irrigated area.
16. The food price rises currently being experienced in China are generally attributed to rising demand, poor weather and outbreaks of livestock disease. Most commentators agree that the current global spike in food prices has little to do with China since the country continues to be largely self-sufficient in grain (Wiggins, 2008).
17. Only in 1998 were responsibilities for groundwater monitoring and management transferred to MWR from the Ministry of Mines (now the Ministry Land Resources—MLR) and the Ministry of Construction (in urban areas). However, most groundwater data and knowledge is still found within the hydrogeological branches of the MLR rather than with the MWR and its subordinate Water Affairs Bureaus (Foster *et al.*, 2004).
18. Under the 2002 Law, all water resources are owned by the state. Although state ownership is a cardinal principle of socialist legality, historically it has not led to effective control. This reflects, in part, China's civil law, allowing subordinate units of government to develop relatively firm entitlements and over-use resources (Wouters *et al.*, 2004).
19. River basin conservancy commissions have been established in six key river basins, including the Yellow River.
20. The conversion of Water Resource Bureaus to Water Affairs Bureaus began in Shenzhen in 1991, and has led to the consolidation of water resources development, management, flood control and rural-urban water supply under one roof (Shah *et al.*, 2004).
21. In practice, annual regulation plans are not prepared for many rivers, particularly in southern China where water resources are more abundant.
22. With certain exemptions, for example for stock and domestic purposes in rural areas, and rural collectives taking water from their own works.
23. Under the 1988 Water Law the state was required to adopt a permit system to regulate direct withdrawals from aquifers, rivers and lakes. In September 1993, the state also issued Implementation Procedures for the Water Drawing Permit System, outlining the scope and implementation measures for the permit system.
24. Permit No. 1 for Hangjin ID allocates 410 million m^3 to the district, including a mandatory return flow of 35 million m^3 per year. Permit Nos. 2 and 3 for Hetao ID allocate a much larger volume of 4.82 billion m^3.
25. Hence membership of the WUA, conferred through village registration and land ownership in the ID, provides farmers with rights to an irrigation service, subject to pre-payment, with accountability provided ultimately through voting rights. However, individual farmers only have an indirect role in ensuring the WUA does not lose any contractual water rights granted to it.
26. This assumes that channel leakage was not being used to maintain environmental assets or provide usable recharge to groundwater users. In many closed basins, this assumption may not be valid (FAO, 2004; Perry, 2007).
27. A typical groundwater district in the US or Australia might include one thousand farmers. In an area of comparable size in China, there may be 100 000 farmers, each withdrawing small volumes of water (Shah *et al.*, 2003).
28. Permits for well drilling can serve to check the numbers of groundwater users as well as the location and spacing of wells. However, abstraction licenses that (ideally) define variable shares of aquifer safe yield are required to control total groundwater withdrawals.

References

Ash, R. F. (1993) Agricultural policy under the impact of reform, in: Y. Y. Kueh & Robert F. Ash (Eds) *Economic Trends in Chinese Agriculture: The Impact of Post-Mao Reforms* (Oxford: Oxford University Press).

Bramall, C. (2000) *Sources of Chinese Economic Growth 1978–96* (Oxford: Oxford University Press).

Brown, L. (1995) *Who Will Feed China? Wake-up Call for a Small Planet* (World Watch Environmental Alert Series) (New York: W.W. Norton).

Calow, R., Wang, Z. & Zheng, H. (2008) Report on Management in Hetao Irrigation District, unpublished report to DEWHA under the WET Program.

Calow, R., Zhao, Y., Anscombe, R., Wang, H. & An, L. (2006) *Managing groundwater in rural Hebei: Lessons from the Water and Agricultural Management in Hebei (WAMH) project* WAMH Project Briefing Paper, based on paper presented at the 34th International Association of Hydrogeologists (IAH) Congress, Beijing, October.

Clayre, A. (1984) *The Heart of the Dragon* (London: W. Collins).

FAO (2004) *Water Charging in Irrigated Agriculture: An Analysis of International Experience,* FAO Water Report No. 28, by G. Cornish, B. Bosworth & C. Perry, with J. Burke (Rome: Food and Agriculture Organisation of the United Nations).

Foster, S., Garduno, H., Evans, R., Olson, D., Tian, Y., Zhang, W. & Han, Z. (2004) Quaternary aquifer of the North China Plain: Assessing and achieving groundwater resource sustainability, *Hydrogeology Journal*, 12, pp. 81–93.

Foster, S. & Garduno, H. (2004) Towards sustainable groundwater use for irrigated agriculture on the North China Plain, in: *GW-MATE Case Profile Collection No.8* (Washington, DC: World Bank).

Lohmar, B., Wang, J., Rozelle, S., Huang, J. & Dawe, D. (2003) China's agricultural water policy reforms: increasing investment, resolving conflicts and revising incentives, *Agricultural Information Bulletin*, No.782, Economic Research Service, U.S. Department of Agriculture.

MWR (2001) *China Agenda for Water Sector Strategy for North China*, Vol. 2, Main Report 2 April 2001, Report No. 22040-CHA (Beijing: Ministry of Water Resources).

MWR (2003) *Advice on Small Rural Water Conservation Reform Implementation* (Beijing: Ministry of Water Resources).

MWR (2006a) *Annual Report 2004–2005* (Beijing: Ministry of Water Resources of the People's Republic of China).

MWR (2006b) *2005 Statistic Bulletin of China's Water Activities* (Beijing: Ministry of Water Resources of the People's Republic of China).

NBSC (2004) *China Development Report 2004* (Beijing: National Bureau of Statistics of China).

NDRC (2000) *Several Suggestions on the Reform of Water Prices Used on Agricultural Problems* (Beijing: National Development and Reform Commission).

OECD (2005) *OECD Review of Agricultural Policies: China 2005* (Paris: OECD).

Perry, C. J. (2007) Efficient irrigation; inefficient communication; flawed recommendations, *Irrigation and Drainage*, 56, pp. 367–378.

Shah, T., Deb Roy, A., Qureshi, A. S. & Wang, J. (2003) Sustaining Asia's groundwater boom: An overview of issues and evidence, *Natural Resources Forum*, 27, pp. 130–141.

Shah, T., Giordano, M. & Wang, J. (2004) Irrigation institutions in a dynamic economy: What is China doing differently to India?, *Economic and Political Weekly*, 39(4), pp. 361–370.

Shen, D. & Speed, R. (2009) Water resources allocation in the People's Republic of China, *International Journal of Water Resources Development*, 25(2), pp. 209–225.

Spence, J. D. (1999) *The Search for Modern China* (New York: W.W. Norton & Company).

Stone, B. (1993) Basic agricultural technology under reform, in: *Economic Trends in Chinese Agriculture: The Impact of Post-Mao Reforms* (Oxford: Oxford University Press).

WAMH (2007) *Water Resources Legislative Review and Related Institutional Arrangements, Hebei*, Final Report for the Water and Agricultural Management in Hebei Project, May 2007, Shijiazhuang, Water and Agricultural Management in Hebei Project Report.

Wang, J., Huang, J., Rozelle, S., Huang, Q. & Blanke, A. (2007a) Agriculture and groundwater development in northern China: Trends, institutional responses and policy options, *Water Policy*, 9(Supplement 1), pp. 61–74.

Wang, J., Huang, J., Blanke, A., Huang, Q. & Rozelle, S. (2007b) The development, challenges and management of groundwater in rural China, in: M. Giordano & K. G. Villholt (Eds) *The Agricultural Groundwater Revolution: Opportunities and Threats to Development*, Comprehensive Assessment of Water Management in Agriculture Series, pp. 37–62 (Trowbridge: Cromwell Press).

Wang, J., Huang, J., Rozelle, S., Huang, Q. & Zhang, L. (2008) Understanding the water crisis in northern China: How do farmers and the government respond? in: Ligang Song & Wing Thye Woo (Eds) *China's Dilemma: Economic Growth, Environment and Climate Change*, pp. 277–296 (ANUE Press, Asia Pacific Press, Brookings Institution Press and Social Sciences Academic Press (China)).

Wang, Z., Hang, Z. & Wang, X. (2009) A harmonious water rights allocation model for Shiyang River Basin, Gansu Province, China, *International Journal of Water Resources Development*, 25(2), pp. 355–372.

WET (2006) *Water Entitlements and Trading Project (WET Phase 1) Final Report* November 2006 [in English and Chinese] (Beijing: Ministry of Water Resources, People's Republic of China and Canberra: Department of Agriculture, Fisheries and Forestry, Australian Government). Available at: http://www.environment.gov.au/water/action/international/wet1.html.

WET (2007) *Water Entitlements and Trading Project (WET Phase 2) Final Report* December 2007 [in English and Chinese] (Beijing: Ministry of Water Resources, People's Republic of China and Canberra: Department of the Environment, Water, Heritage and the Arts, Australian Government). Available at: http://www.environment.gov.au/water/action/international/wet2.html

Wiggins, S. (2008) Is the global food system broken? *ODI Opinion*, 113(October) (London: Overseas Development Institute).

World Bank (2007a) *Cost of Pollution in China: Economic Estimates of Physical Damages* (Washington, DC: World Bank).

World Bank (2007b) *Implementation Completion and Results Report, Water Conservation Project* Report No. ICR0000191 (Washington DC: World Bank).

Wouters, P., Hu, D., Zhang, J., Tarlock, D. & Andrews-Speed, P. (2004) The new development of water law in China, *University of Denver Water Law Review*, 7(2), Spring 2004, pp. 243–308.

Urban Water Management in China

MARTIN COSIER & DAJUN SHEN

ABSTRACT   *This paper summarizes the urban water management framework that China's central government has adopted to manage the supply of water to the country's urban areas. These laws, policies and institutional arrangements cover water resources management, urban planning and environmental management issues. This paper considers the application of this framework in three urban areas: Beijing, Shanghai and Shaoxing Prefecture in Zhejiang Province. These case studies are used as the basis of a discussion on the ways that the framework could be enhanced, both structurally and in terms of its implementation.*

Introduction

Urban water supply presents challenges that are unique in water resource management. These primarily relate to the need for very high reliability and security of supply and the mix of social and economic needs for water. Simply put, urban communities increasingly expect a constant supply of high-quality water and are less willing (or able, in the case of essential services such as hospitals) to accept supply interruptions or restrictions. Despite these special characteristics, urban water management systems still need to be established and operated within the broader resource management context and be consistent with catchment or regional water resources planning objectives. Ultimately, urban water management is about ensuring the most appropriate allocation and use of a limited natural resource within a framework of economic objectives, social needs and environmental sustainability.

Internationally, there has been recognition through the Millennium Development Goals of the vital importance of improving the provision of safe water supply and sanitation services to the world's poor (UN, 2008). These objectives, together with the ever-growing challenges associated with urbanization and climate change, have only served to heighten pressure to improve urban water management and supply arrangements. In managing urban water supply, jurisdictions around the world have adopted various institutional arrangements, regulatory regimes and supply planning approaches. Governments tend to take a lead role in water supply due to the monopolistic and essential nature of the sector (in fact, in South Africa legislation requires that all households receive 200 litres of water per day for free; Kessides, 2004). As urban water supply and wastewater disposal systems also have common infrastructure requirements—supply sources, raw water treatment plants and distribution infrastructure, wastewater collection, treatment and disposal facilities (IPART, 2007)—it is most common to find vertically integrated monopolists, responsible for supply, distribution and retail services (Kessides, 2004). Opportunities for competition have therefore been limited and commercial involvement in the urban water sector has required initiation by governments.

A wide range of approaches involving the private sector in the ownership and financing of urban water supply systems has been used around the world. This has ranged from the private operation of publicly owned and funded systems via service contracts (common in France) to joint public-private operation and ownership via joint ventures (used in Colombia) and full privatization (such as in Great Britain). There has also been an increase in the corporatization of the management of publicly owned utilities (OECD, 2000). Other notable trends in urban water management relate to mechanisms to provide supply reliability in the face of resource scarcity. For example, desalination is increasingly being used in Australia, the Middle East and the United States. Australia, Singapore and the United States are also reconsidering traditional approaches to wastewater services, by treating stormwater and wastewater streams as valuable resources for reuse (IPART, 2007).

China's water resource management system reflects the country's approach of having policies and laws set centrally, but administered locally. This system combines the multi-level jurisdictional framework (national, provincial, prefecture and county levels of administration) with a catchment-based approach to river basin management (Shen & Liu, 2008). The associated management arrangements are costly due to their complexity, the need for a high level of consistency and the involvement of multiple government agencies, each with their own priorities. When combined with the need to provide highly reliable supplies to a diverse range of users, these factors mean urban water management is arguably the most challenging component of water resource management in China.

Research on urban water management systems in China has tended to focus on private sector financing arrangements for urban supply networks and other specific management aspects, such as pollution control or increased reuse of wastewater (for example, see Zhong *et al.*, 2008; Lee, 2006; Yang & Abbaspour, 2007). More comprehensive studies that also consider broad governance arrangements, such as the recent review of urban water utility performance by the World Bank (2007), have not focussed on the interaction of the urban water management systems with broader water resource management arrangements.

In summarizing the laws, policies and institutional arrangements that China has adopted at the national level to manage the supply of water to its urban areas, this paper focusses on water resources management, urban planning and environmental protection and how they fit into the overall management framework. This is followed by case studies in three urban areas: Beijing, Shanghai and Shaoxing prefecture in Zhejiang Province (Figure 1). It then discusses a range of ways that China's urban water management framework could be improved, both from a structural perspective and in terms of its implementation.

Overview of China's Urban Water Management Framework

China's urban water management arrangements are primarily set through national laws and policies, for implementation by provincial, prefectural and county governments. Generally, the institutional division of responsibilities at the national level is reflected in equivalent line agencies at each of the lower levels of government (Shen & Liu, 2008). National laws and

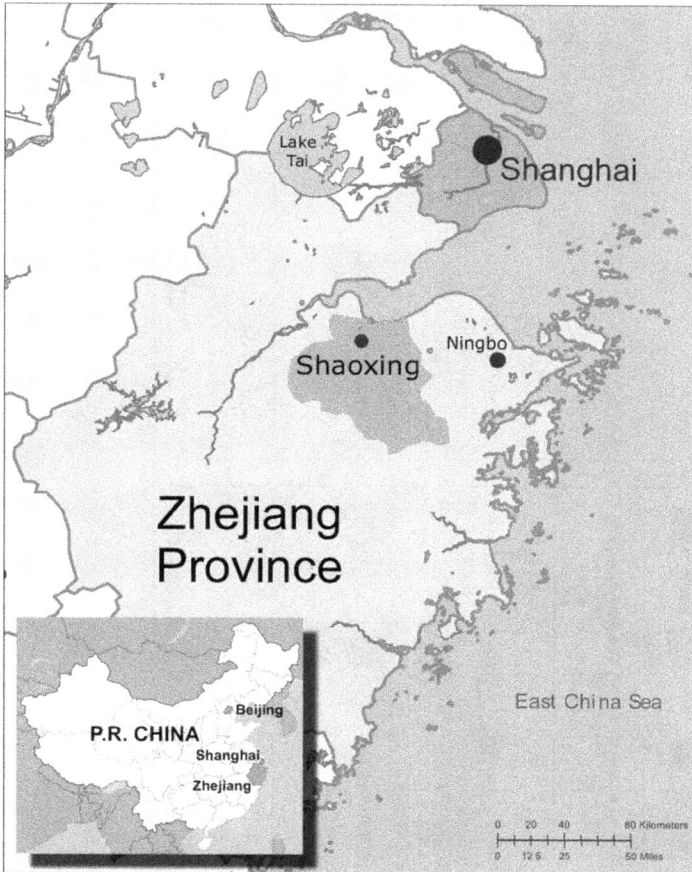

Figure 1. Maps of case study areas.

policies are formulated as part of broader arrangements relating to water resources management, urban planning and environmental management. While these three policy areas sometimes have differing objectives in relation to urban water issues (for example, urban planning concerns to meet future growth in demand do not necessarily consider the environmental implications of continuous development), there are some mechanisms designed to align the various tools into an urban water management framework. This framework is depicted in Figure 2 and summarized in the following paragraphs.

The water resources management arrangements that influence the urban sector are largely provided by the 2002 Water Law and are primarily the responsibility of the Ministry of Water Resources. This legislation requires the preparation of 'water resources allocation plans' to determine the water that is available for different consumptive purposes within a river basin and within administrative regions in that basin. Availability of water is determined on a yearly basis through 'annual water resources regulation plans'. Water abstraction by different users is governed by a 'water abstraction permit system', provided for by the 2002 Water Law and the 2006 Regulations on the Administration of Water

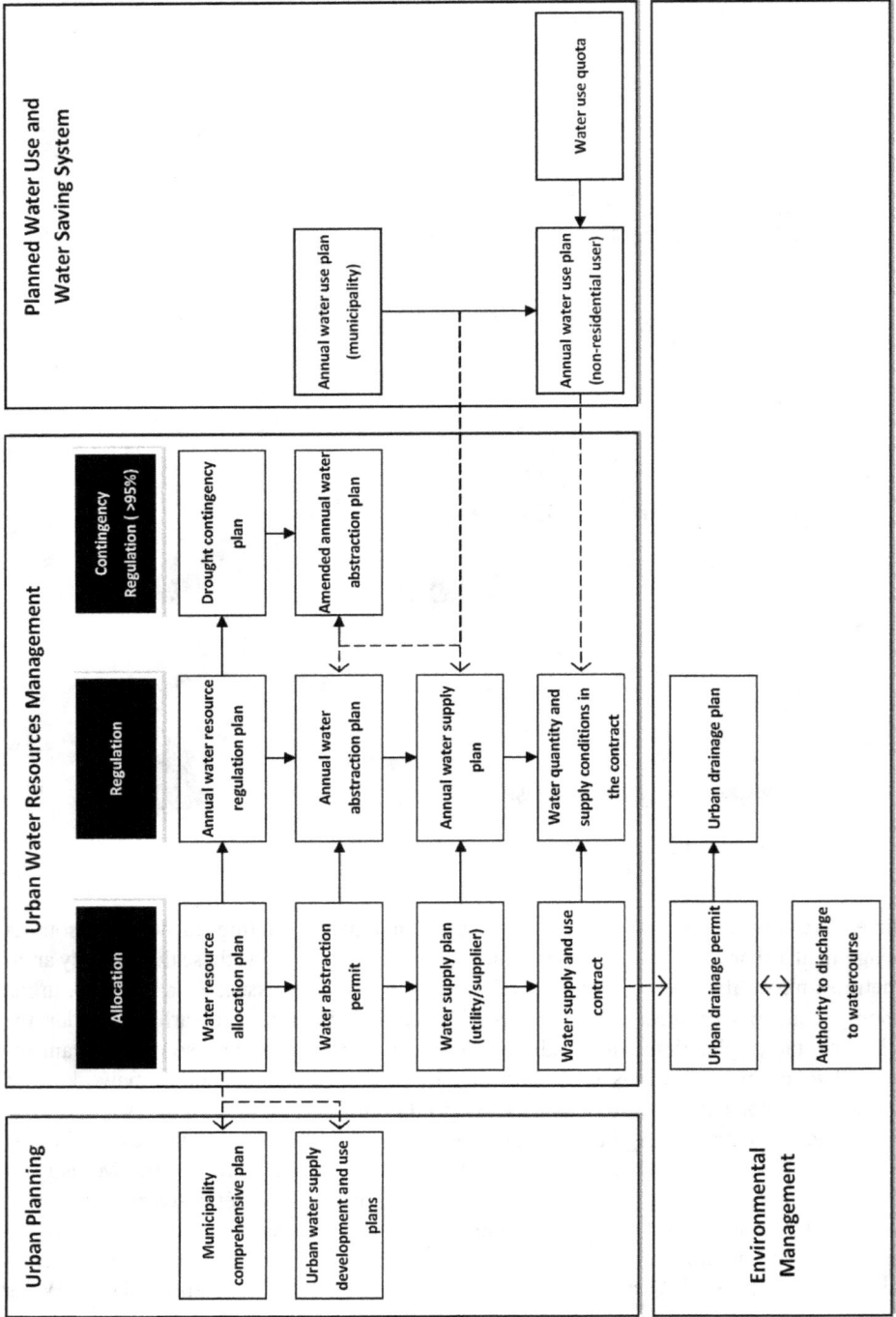

Figure 2. China's urban water supply and management framework, grouped by framework components.

Abstraction Licensing and Collection of Water Resources Charges. Permits are regulated on an annual basis through 'annual water abstraction plans', which must be consistent with relevant annual water resources regulation plans. In periods of extreme drought, 'drought contingency plans' take precedence over other plans and require amendments to annual water abstraction plans. These contingency arrangements are primarily designed to protect the availability of water for urban users, particularly for essential domestic needs, which are given priority under the 2002 Water Law. (For a detailed explanation of the water allocation and regulation process, see Shen & Speed, 2009.)

In the urban context, permits are generally held by public water supply entities (sometimes bulk suppliers and sometimes retail utility companies) that provide urban water supply services to the end users (although some major industrial users operate their own water abstraction and supply systems). The rights and responsibilities of the suppliers and users in relation to the ongoing and annual entitlements to water are governed both by 'water supply and use contracts' between these parties and by the suppliers' 'water supply plans'. These plans and contracts cover matters such as the price paid for water services, the rate and pressure of water supply, metering obligations and standards for meters, and the quality of water supplied. The Ministry of Health sets national drinking water quality standards that must be met by urban water suppliers. These mechanisms are required and managed under the 1994 Urban Water Supply Regulations, 1989 Urban Planning Law, 1999 Contract Law, 2006 Regulations on the Administration of Water Abstraction Licensing and Collection of Water Resources Charges, and 2007 Urban Water Supply Quality Regulations.

The use of water in China's cities is subject to a 'planned water use and water saving system' under the national 1988 Regulations on Urban Water Saving Management, which is designed to minimize wastage of water and to increase water use efficiency. This system requires the preparation of 'annual water use plans' for the city as a whole, as well as for individual non-residential users. Municipality annual water use plans define the way that water available to the municipality each year is to be used. Non-residential users' annual water use plans effectively provide a yearly allowance. These allowances are determined by reference to 'water use quotas' for user classes in the region that are intended to normalize water use within particular industries—the quotas are set for a single unit of production or service for similar users in a region. Higher water charges are imposed, using inclining block tariffs, on users that exceed the allowance under their water use plan.

Urban water pricing is regulated at the national level by the National Development and Reform Commission. While residential users are not captured by this system, the 2002 Water Law requires local municipal governments to improve domestic water efficiency by, for example, promoting the use of water efficient appliances within households, reducing leakage from supply networks and increasing the use of recycled water.

Urban planning is highly important in the context of China's continued urbanization, rapid economic growth and associated increases in living standards. State Council approved 'municipal comprehensive plans' are required for significant expansions of existing urban areas or new economic development zones (1989 Urban Planning Law). In line with these plans and under the 1994 Urban Water Supply Regulations, municipalities are required to prepare medium- to long-term plans for the development of water sources to supply the growth in urban demand. While these 'urban water supply development and use plans' are concerned with meeting future urban water needs, their preparation needs to be coordinated with the water allocation planning framework to ensure the sustainable use of available resources.

The Ministry of Housing and Urban-Rural Development (the Ministry of Construction prior to the 2008 governmental restructure) and its local equivalents (urban construction departments) are technically responsible for the development of urban water sources and delivery networks under the 1994 Urban Water Supply Regulation. However, local water resources departments have tended to take an active role in urban development planning. Recently, the broad responsibilities for the development of urban water supplies have been formally transferred to the water resources departments in some parts of the country where urban and rural water management is being integrated (Shen & Liu, 2008).

Wastewater discharge is managed through an 'urban drainage permit system' under the national 2007 Methods on Urban Drainage Permit Management. This system uses urban drainage permits and urban drainage plans to regulate the quality, quantity and flow of wastewater that enters distribution and treatment infrastructure. Local drainage departments are required to regularly test wastewater releases to ensure that the quality meets the requirements under the national 1996 Comprehensive Standard of Sewage Drainage—lack of compliance can lead to cancellation of the drainage permit. In some situations, urban users are authorized to discharge wastewater directly to watercourses. These situations are assessed on a case-by-case basis and require approval under a separate permit system from the local environmental protection department (2008 Law on the Prevention and Control of Water Pollution). However, the distinction between the two systems is not entirely clear. In recognition of overlaps, the 2008 government restructure resulted in certain information-sharing requirements between the Ministry of Water Resources and the Ministry of Environmental Protection. In particular, the agencies are required to agree on any new water quality standards or information to be publicly released to ensure there is a single government data set.

Urban Water Supply in Beijing

China's capital city is a centrally administered municipality with an official population of 14.2 million and approximately four million additional migrant workers. It covers almost 17 000 km^2, about 1000 km^2 of which is the urban footprint. The municipal government is responsible for 16 urban districts and two rural counties (Beijing Municipal Government, n.d.). The municipality is surrounded by Hebei Province and sits within the Haihe (Hai) River Basin. Beijing's annual water resource availability is estimated at 4.1 billion m^3. Two-thirds of Beijing's supply is from groundwater found on the alluvial fans of the Yongding and Chaobai Rivers, which has been significantly overexploited since the 1960s. However, the bulk of the groundwater use is for agriculture, which accounts for almost half of Beijing's total use (Wei, 2005). The major surface water sources for urban use are the Miyun and Huairou Reservoirs. To help address ongoing water shortages, the city's supplies will be supplemented by water diverted from the Yangtze River Basin as part of the North-South Water Transfer Project. From 2010, one billion m^3 is planned to be diverted annually (Wei, 2005).

Institutional Arrangements

The Beijing Water Affairs Bureau, an administrative unit of the Beijing Municipal Government, is chiefly responsible for both urban and rural water management, including water abstractions, supply planning, water use efficiency and wastewater discharge.

It comprises a number of departments responsible for specific water management functions, including the Water Resources Division, the Water Supply Management Division and the Water Saving Office. Various affiliated agencies, such as the Beijing Water Saving Management Centre, support specific aspects of the bureau's functions. Responsibilities for implementing the planned water use system and water saving system in Beijing are theoretically split between two main parts of the Water Affairs Bureau. The Water Saving Office is responsible for setting policy directions to be implemented by the Beijing Water Management Centre (2005 Water Saving Methods). In practice, however, there is some overlap and, at times, reassignment of the work of these two divisions.

A number of other agencies within the Beijing Municipal Government participate in managing the city's urban water supplies. The Beijing Municipal Commission of Development and Reform determines the city's economic and social development strategies, which includes overseeing water charges. The Urban Water Authority within the municipality's construction department undertakes annual water use planning for non-residential urban users, but its role is not as significant as elsewhere in China because most management functions have been transferred to the Beijing Water Affairs Bureau.

Beijing's water supply and wastewater disposal networks are operated by separate corporate entities that are wholly owned by the Beijing Municipal Government. The Beijing Waterworks Group Company supplies water to 100% of Beijing's urban population through six branch corporations. At the end of 2006, this company operated 19 waterworks plants, with an aggregate daily water supply capacity of 2.93 million m^3 (700 million m^3 per year). Its distribution network is over 8000 kilometres long and covers more than $600 km^2$. Each waterworks plant produces water independently, but these supplies are controlled by the Beijing Waterworks Group Company's dispatch centre.

The Beijing Waterworks Group Company's branch corporations are named after particular treatment plants, despite multiple plants being managed under each branch. The largest is the Ninth Waterworks, which supplies about 50–60% of the municipality's total urban water demand from the Miyun and Huairou Reservoirs. It has a daily water supply capacity of 1.5 million m^3. In recent times, surface water shortages have led the Ninth Waterworks to access emergency groundwater supplies of up to 500 000 m^3 per day. Water diverted from the Yangtze River Basin via the North-South Water Transfer Project will be managed by the Ninth Waterworks (Water Entitlements and Trading Project (WET), 2007). Smaller volumes of surface water from the Miyun Reservoir are treated by another branch corporation, the Tiancunshan Water Treatment Plant. The Third Waterworks, Fourth Waterworks and Eighth Waterworks corporations supply other parts of Beijing's urban areas with groundwater. While the water treatment plants produced water that met both World Health Organization and Chinese national standards for drinking water quality by the start of the 2008 Olympics, much of the city's distribution network comprises very old pipes that re-contaminate this water on route to users (Liu, 2007, 3 July; UNEP, n.d.).

The Beijing Drainage Group Company manages the wastewater drainage network, which has a total length of nearly 4000 kilometres. The company operates 11 mid- to large-scale municipal wastewater treatment plants that have a combined daily treatment capacity of 2.8 million m^3, 2.56 million m^3 of which is in Beijing's urban areas (Beijing Drainage Group Company, n.d.). This represents a significant increase from the 1.09 million m^3 that could be treated daily in 1999 and is in line with commitments made by the Beijing Municipal Government as part of the bid for the 2008 Olympics. The proportion of

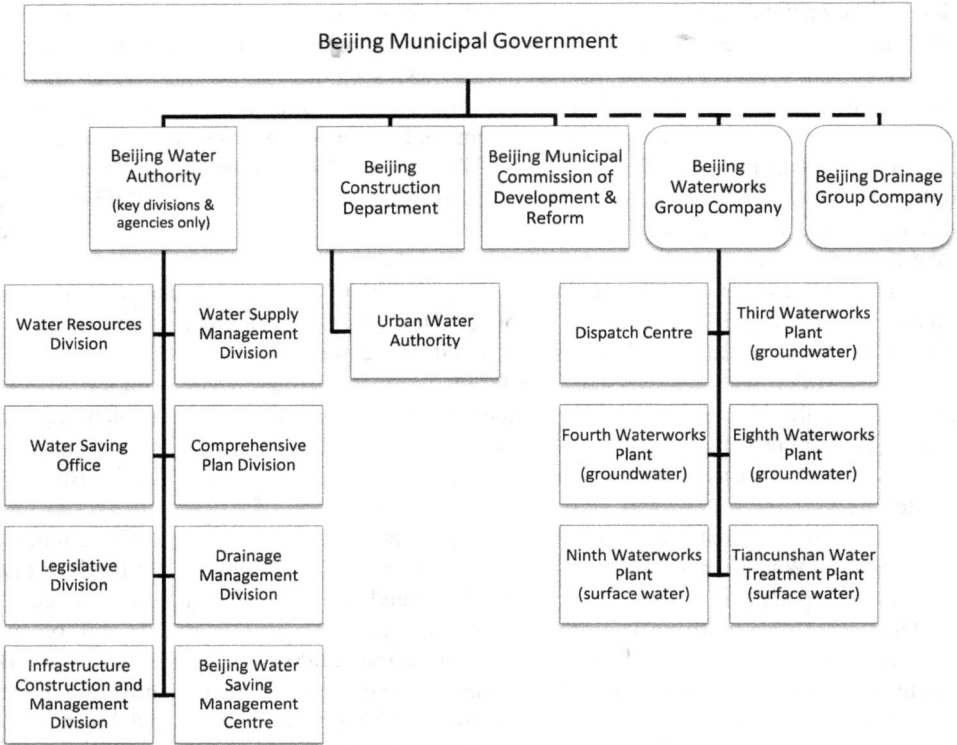

Figure 3. Urban water supply and management institutional arrangements in Beijing. *Notes:* 1. The Beijing Waterworks Group and the Beijing Drainage Group are independent entities owned by the Beijing Municipal Government. 2. The Beijing Water Saving Management Centre is affiliated to the Beijing Water Authority, but is not a government division.

urban wastewater treated each year has also risen drastically, from 22% in 1998 to 90% in 2006 (UNEP, n.d.). Beijing's urban water supply and management institutional arrangements are depicted in Figure 3.

Supply and Management Practices

In 2001, the State Council approved the "Beijing Water Resources Sustainable Use Plan for the Early 21st Century" followed in 2007 by the "Yongding River Trunk Stream Water Resources Allocation Plan". Together, these documents act as water resources allocation plans for Beijing by defining the municipality's regional water allocation in terms of the flows that provinces upstream of Beijing are required to maintain in the rivers that feed Beijing's reservoirs. In the medium- to long-term this allocation will not be able to meet the city's demands, hence the major undertakings to find new supplies such as the North-South Water Transfer Project.

The Water Resources Management Division in the Beijing Water Affairs Bureau is responsible for the management of water abstraction permits. The branch corporations under the Beijing Waterworks Group Company hold water abstraction permits for the city's urban supplies. The waterworks plants were constructed before the water abstraction permit system was implemented, while the permits were issued before there was a clear definition of Beijing's regional water allocation. In this context, the volumes authorized under the permits were determined according to the size of each pre-existing waterworks plant. As such, the volume of water treated by the waterworks plants cannot (physically) exceed the quantity authorized on the permit. None of the waterworks plants operate under annual water abstraction plans. There is therefore no formal moderation of the year-to-year abstraction of water on the basis of actual conditions.

The Water Resources Management Division also prepares the water supply plan for the Beijing Waterworks Group Company. In practice, however, the company's day-to-day operations simply involve the dispatch centre estimating the coming day's demand and directing the branch corporations to release certain volumes of water accordingly. All the waterworks plants have been linked and are managed as a single network to guarantee the water demands of users.

The Beijing Water Saving Management Centre manages the planned water use and water saving system in accordance with the national 1988 Regulations on Urban Water Saving Management and the Beijing 2005 Water Saving Methods. Under this system, the annual use of water by individual non-residential users is controlled via 'water use indexes' (the city's term for water use plans). The individual allowance provided by the water use index is based on the municipality's annual water use plan, the relevant sectoral water use quota and general social and economic objectives. About 30 000 indexes are managed under the system. Often the indexes remain constant, but the centre also considers applications for increased allowances and from new users. The annual water use index is broken down into monthly intervals. Higher water tariffs are imposed for use in excess of the monthly volumes. The water use index system also incorporates mandatory water saving requirements imposed by the Beijing Municipal Government. Current water savings efforts are targeted at government agencies and high-consumption industries.

The primary considerations in preparing an annual water use index are the relevant water use quota and the user's past consumption levels. The municipality's abstraction caps under the water abstraction permits are not formally considered in setting the annual allowances for users. This could be due to the city's overall consumption not being able to exceed the abstraction caps, given that they are set at the physical limits of the treatment plants. Regardless of the reason, this means that water managers are unlikely to consider the consequences of future changes to one—abstraction levels or water use indexes—on the other.

Beijing's long-term water supply strategy includes increasing the use of recycled water. There are plans for urban areas to have 14 new wastewater treatment plants with a total daily capacity of 3.25 million m^3 in order to achieve an annual recycled water production target of 640 million m^3 by 2010 (Wei, 2005). Only about 20% of treated wastewater is currently reused in Beijing (Yang & Abbaspour, 2007). The national drainage permit system has not been comprehensively implemented in Beijing. As such, the quantity and quality of wastewater discharged to the urban drainage network is not well regulated. There is, however, some regulation of the discharge of pollutants into the natural environment by the municipal environmental protection bureau (Ma & Ortolano, 2000).

Urban Water Supply in Shanghai

Shanghai, like Beijing, is a centrally administered municipality. It is located on China's east coast at the mouth of the Yangtze River and lies between Jiangsu and Zhejiang Provinces. About 1000 kilometres south of Beijing, Shanghai has a subtropical climate. At the end of 2005, the city's population was officially estimated at 17.8 million (Shanghai Municipal Government, 2007), although the actual population is likely to be well over 20 million (*China Daily Online*, 2003). Shanghai is the most densely populated of China's administrative regions, with an average of 2145 people per km^2 (Shanghai Municipal Government, 2007; Shen & Liang, 2003). It incorporates 19 districts and one county that collectively cover 6340.5 km^2 (Shanghai Municipal Government, 2007), 140 km^2 of which is classified as urban (Lee, 2006).

Shanghai's water is sourced from the Yangtze River and Tai Lake Basins. Its two main urban sources—the Huangpu River, which flows through the middle of the municipality, and the Yangtze River—together supply about 2.5 billion m^3 per year. Only limited supplies are drawn from Shanghai's groundwater sources, which have been significantly over-extracted in past years, causing land subsidence of between 1.5 and 2.5 metres in parts of the city (OECD, 2007). In 2007, the peak daily water supply in Shanghai was 10.25 million m^3. About 80% of the city's total surface water supplies are currently sourced from the Huangpu River. However, a new offtake from the Yangtze River is being constructed, which will take the total water abstraction capacity from the river to 7 million m^3 per day. Due to its superior water quality, following the completion of these new works the level of abstraction from the Yangtze River is planned to be increased in stages until it reaches 70% of the city's supplies. Shanghai has 123 water treatment plants but many of these do not produce water of an acceptable quality and will be decommissioned over the coming years. These projects reflect the Shanghai Municipal Government's view that the city has sufficient water resources, but that there is a need to improve the quality of water supplied to its users, particularly for drinking purposes.

Institutional Arrangements

The Water Affairs Bureau in Shanghai undertakes the overall management of water affairs in both the city's urban and rural areas. The Bureau's responsibilities for resources management, supply security and environmental management issues are grouped into five function areas: water supply for urban and rural areas; drainage and wastewater treatment; management of groundwater; planned water use and water saving; and flood prevention. The policy formulation, policy implementation, and enforcement and support aspects of these function areas have been institutionally separated (depicted in Figure 4).

The Bureau's Water Resources Management Department is responsible for preparing water management policies relating to the: supervision of water abstraction permits; collection of water resource fees; management of the water supply sector; planned water use and water saving system; supervision of drainage and sewage treatment; protection of water source areas; hydrological and water quality monitoring systems; and management of groundwater. Several agencies are responsible for the implementation of these policies. The Water Supply Management Department (incorporating the regulation and monitoring centre) is responsible for implementing the water abstraction permit system, collecting water resource fees, overseeing waterworks enterprises and undertaking water efficiency

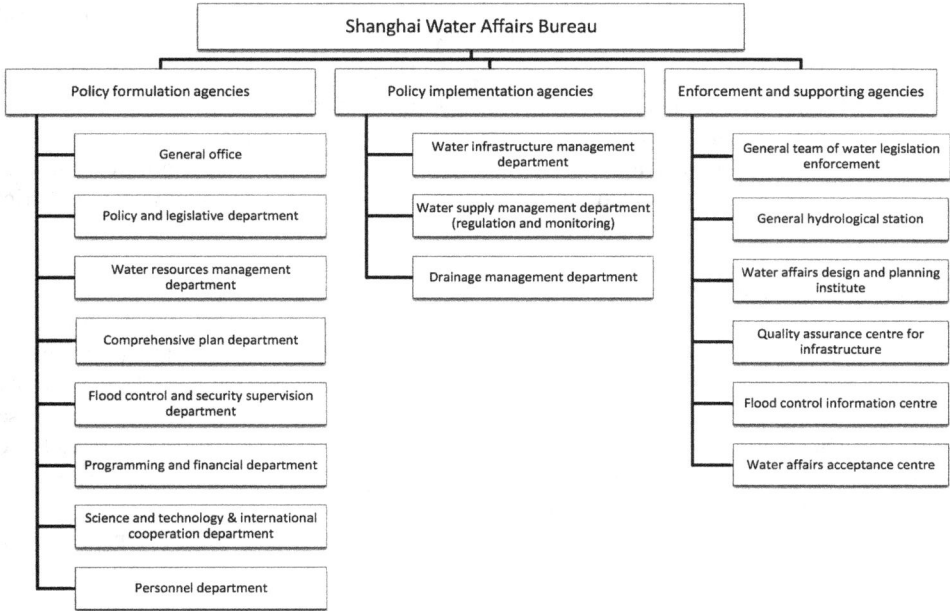

Figure 4. Structure of the Shanghai Water Affairs Bureau.

measures. The Drainage Management Department manages the daily operations of the drainage network and wastewater treatment plants. The Water Infrastructure Management Department oversees small scale irrigation engineering, watercourse management and sectoral management of floodgates and dikes.

Separate corporations operate the two raw water supply plants that abstract water from the city's supply sources (Shen & Liang, 2003). These raw water supply corporations supply four water utility companies that operate the city's water treatment plants and provide urban water supply services within defined geographical management districts. One of these utilities, the Shanghai Pudong Veolia Water Company, is operated as a Chinese-foreign joint venture. In 2002, following a competitive tender process, 50% of the state's equity in the utility company was transferred to France's Veolia Group. This was the first such foreign investment arrangement in China's water sector. The joint venture operates under a concession contract with a 50-year term from the Shanghai Municipal Government (Fu *et al.*, 2008). Shanghai also has three drainage companies with responsibility for distinct geographical regions and two large wastewater treatment plants.

Supply and Management Practices

Water resources allocation plans have not yet been prepared for the Yangtze River or Tai Lake Basins. However, a comprehensive plan for Shanghai is currently being drafted for State Council consideration and will include information about water availability for the municipality. This plan will essentially provide a regional water allocation for Shanghai and, amongst other provisions, is likely to include controls on the use of water from Tai Lake, including setting minimum inflow rates and water levels in the lake that will need to be maintained before any abstractions can occur. This comprehensive plan will also form the basis for preparing annual water resources regulation plans for the city.

The Shanghai Water Affairs Bureau's Water Supply Management Department is responsible for considering applications for new water abstraction permits for volumes greater than 20 000 m^3. Applications for smaller abstractions are considered at the district level. The factors considered when examining and approving applications for water abstraction permits for urban purposes are the 'water resources justification report', which must demonstrate that the urban supply reliability will be at least 99%, and the general circumstances of the intended water supply area, such as the area's population and industry levels. The water abstraction permits for Shanghai's reticulated urban water supplies are held by the two raw water supply corporations. Each year, the Water Supply Management Department prepares annual water abstraction plans for each of these permits. The authorized abstraction volumes under these plans generally reflect the quantities listed in the permits, but the department also considers the quantity of water that was actually abstracted in the previous year.

Shanghai's four water utility companies prepare annual water supply plans based on the water demand within their region of operation, long-range weather forecasts and local development plans (based on dialogues with the municipal urban construction department). The water use in relatively developed areas is stable, but consumption levels are increasing in suburban areas. As such, the degree to which these plans change on an annual basis varies between the different water utilities. Generally, these annual water supply plans are broken down into quarterly and monthly plans, which guide short-term water supply but do not act as periodic limits. The plans are prepared in consultation with the raw water supply corporations and are incorporated into their water supply contracts with the water utilities. These contracts, which are overseen by the Water Affairs Bureau, are agreed at the beginning of each year and govern the supply of water to each treatment plant (in accordance with the utilities' annual water supply plans) and the price paid to the raw water supply corporations for these services. The water supply contracts are not, however, formally coordinated with the annual water abstraction plans that govern the volume of water available in a given year to the raw water supply corporations. While the water utilities provide reticulated supply to 99.9% of Shanghai's urban users, these water supply services are not governed by contracts between the water utilities and the end users.

The regulation and monitoring centre within the Water Supply Management Department regulates the daily supply of water for the municipality's urban users. Generally, the centre holds monthly water regulation meetings with the raw water supply corporations and the water utilities to determine a water regulation plan for the following month. The centre regulates real-time water supply according to the water pressure in the networks.

Shanghai's 1994 Management Methods on Water Saving governs the implementation of the planned water use and water saving system. About 18 000 users have annual water use plans that provide monthly allowances—users that exceed these allowances are charged penalty rates. These plans are generally carried over from year to year unless the water user seeks an increased allowance. Major users have their water use plan, actual consumption levels and water efficiency measures reassessed every five years. Where such a review finds opportunities for greater water savings, the user will be subject to on-site regulation to ensure efficiency measures are adopted. Less than 100 new or revised water use plans are issued each year by the Water Supply Management Department.

At present, hotels, schools and hospitals are under 'complete quota management', whereby these entities' water use plans are defined by the sector's water use quota. The quotas have been set to provide an acceptable level of water use per unit of service for each of these sectors, which is then up-scaled to determine the particular user's annual allowance. The Water Affairs Bureau plans to incorporate additional sectors into its complete quota management system, starting with the power and steel industries. Until then, water use plans for users that are not yet subject to quotas are prepared following an assessment of the particular user's demand, actual usage in past years and efforts to increase its water use efficiency.

Since 1990, the Shanghai Water Affairs Bureau has operated a drainage permit system, consistent with national requirements, to manage wastewater discharge to the urban public sewerage network and wastewater treatment system. The permits authorize the holders to discharge wastewater in accordance with quality standards set by the Shanghai Municipal Government. The system includes the collection of wastewater discharge fees from permit holders and penalty fees where the discharged water breaches required standards. The revenue from these fees is intended to be used for pollution control activities and to fund loans to industrial water users to establish on-site treatment facilities. However, the fees collected are too low to entirely support these activities and many users remain outside the drainage permit system (Lee, 2006). In 2007, 10 000 users were covered by this system and about 1000 new users are included every year. Some users' permits authorize the discharge of wastewater directly into rivers. In these cases, the decision is made in conjunction with the municipal Environmental Protection Agency, which will issue the permit following an environmental assessment to ensure water quality is not adversely affected and subject to the Water Affairs Bureau approving the discharge outlet.

Urban Water Supply in Shaoxing

Shaoxing is one of eleven prefectures in Zhejiang Province and has a total land area of 8257 km^2. Approximately 230 kilometres to the southwest of Shanghai, it is on the southern edge of the Yangtze River delta and sits on the southern bank of the Qiantang River. Shaoxing incorporates six county-level administrative areas, three of which comprise the prefecture's urban area—Yuecheng District (often referred to as the prefecture's downtown), Shaoxing County (about 15% of the county's population is urban) and Shangyu County. The population of the entire prefecture is 4.36 million, while the population of the combined urban area is about 660 000 (Shaoxing Municipal Government, 2005).

Shaoxing Prefecture receives average annual rainfall of 1448 millimetres and has total water resources of 6.22 billion m^3. While 3.51 billion m^3 of this water has been determined to be available for use on an average annual basis, this has not been formalized through a water resources allocation plan. The main urban water supply source for Shaoxing Prefecture is the Tangpu Reservoir, which has a water supply capacity of one million m^3 per day. Since 2003, when much of Zhejiang Province experienced a significant drought, the total demand for water has not exceeded the supplies available from the Tangpu

Reservoir. The growth in demand, however, is expected to render the current supply capacity insufficient by about 2010.

Institutional Arrangements

The Shaoxing Water Affairs Bureau does not play a major role in managing the prefecture's urban water system because the provincial government is responsible for setting policies governing the implementation of the national management framework and, unlike other parts of China, county-level administrations in Zhejiang have a large implementation role. The Bureau has been corporatized and operates as the Shaoxing Water Affairs Corporation. The corporation's most active policy areas relate to water use monitoring and demand assessments.

Shaoxing's urban water services are operated by separate companies that fall under the Shaoxing Water Affairs Corporation. One of these companies is responsible for the operation of the reservoir, including its offtake works. A second, the Water Treatment Company, controls eight water treatment plants and the main urban supply networks, while a third company owns the distribution network and provides retail water services. Wastewater collection, treatment and disposal services are separated between the Drainage Company and the Wastewater Treatment Developing Company.

The Tangpu Reservoir is jointly owned by the Yuecheng District, Shaoxing County and Shangyu County governments. Their shares of investment in the reservoir were 34%, 33% and 33% respectively. However, the rights under the investment agreement do not correspond to these contributions, with 60% of the water supplied by the reservoir allocated jointly to Yuecheng District and Shaoxing County and 40% to Shangyu County. In practice, though, the water supply situation is different again. In 2007, Shaoxing County received 82 million m^3 from the reservoir, or 43% of the total volume supplied for urban purposes, while Yuecheng District received 70 million m^3 (36%) and Shangyu County received only 40 million m^3 (21%). Subsequent to the original investment agreement, the prefectural government entered into a water rights transfer contract to sell 200 000 m^3 per day to Cixi County in Ningbo (a neighbouring prefecture also in Zhejiang Province). In 2006 and 2007, Cixi received its full entitlement under the contract and the arrangement is expected to continue.

Supply and Management Practices

The management arrangements for the abstraction of water from the Tangpu Reservoir are complicated. The county-level administrations issued themselves with water abstraction permits for their urban supplies, prior to the Zhejiang Provincial Government adopting the national-level water abstraction permit system through the 2006 Details on Implementing Water Abstraction Permits and the 2007 Water Resources Fee Collection Management Methods. These documents require water abstraction permits of the nature held by the county-level administrations to be issued by the Zhejiang Water Resources Department. The provincial water resources department is therefore undertaking a transitional process of issuing three water abstraction permits for the Tangpu Reservoir and, subsequently, will prepare annual water abstraction plans. The volumes under the new permits are expected to reflect the investment contributions each of the county-level administrative areas made towards the development of the reservoir. While the existing entitlements (which will be cancelled) were held by the Water Treatment Company, a decision is yet to be made about which entities will hold the new water abstraction permits.

The Shaoxing Water Affairs Corporation projects both the urban water demand and supply availability on an annual basis. The demand projections are based on past consumption figures provided by the local water treatment plants and include expected peak demands, while the supply estimates take into account predicted reservoir inflows. Water supply, metering and payment arrangements between the companies under the Shaoxing Water Affairs Corporation are managed contractually.

Water supply and use contracts exist between the Shaoxing Water Affairs Corporation and all non-residential users in the urban areas. Contracts cover about 85% of the prefecture's urban residences, while 100% of households have water meters installed. In total, approximately 170 000 contracts were in place by early 2008. The water supply and use contracts with Shaoxing's residential users include provisions about inclining block tariffs. Water charges increase for usage that exceeds $20\,m^3$ per month per household. Average household usage is well below this mark, however, at approximately $7\,m^3$ per month (equivalent to about 78 litres per person per day for a three-person household). Only 5% of users exceed $20\,m^3$ per month. Shaoxing's non-residential urban users do not currently have their annual use controlled under the planned water saving system, as required by national regulations. The prefecture has not required the preparation of water use plans (or adopted any other business efficiency measures) for these users because it is concerned that they will be counter-productive—it considers that current water resource availability is sufficient to supply all urban requirements and that an annual use allowance in a water use plan could be a disincentive to increasing water use efficiency by acting as a target for water users, potentially increasing total usage.

The Drainage Company owns and operates the urban discharge network in Shaoxing, while the wastewater treatment plants are owned and operated by the Wastewater Treatment Developing Company. The relationship between the two entities is governed contractually.

Discussion

The discussion that follows considers the potential for improvements to China's urban water management framework, both in terms of the structure of the framework and approaches to its implementation. Major reform is not proposed, nor indeed considered necessary, but certain areas of the framework may benefit from greater levels of coordination, more aggressive implementation and/or alternative approaches under particular circumstances.

The Structure of China's Urban Water Management Framework

Chinese public policy has a continuing history of establishing independent management systems with limited coordination of their related functions. This has certainly been the case in the water sector. There is obviously a common thread between water abstraction, supply, use (including efficiency measures), quality protection and wastewater disposal, yet the management systems adopted for each of these aspects are not well integrated in practice, resulting in problems for both water resource management generally and urban water management specifically. While the framework—as shown in Figure 2—recognizes the interactions between the different management aspects, the extent of these linkages are not always as strong as might be warranted.

For example, there is no formal mechanism to ensure that the water available to a water utility under a water supply plan is consistent with the volume available—in a resource management sense—under the water abstraction plan. Similarly, there is no legal requirement for annual water abstraction plans to be considered in the preparation of contracts governing the supply arrangements between raw water source suppliers and water utilities or end users. This means that there is no guarantee that the sum of the volumes under all water supply and use contracts (allowing for distribution losses) will be consistent with the water quantity approved in the water abstraction permit. In Shanghai, increased coordination between the water savings and water supply functions could improve the efficiency and effectiveness of both—for instance, the non-residential users' annual water use plans could provide the water utility companies with useful information about industrial water demands for incorporation into their annual water supply plans.

This poor coordination can be attributed to two factors common in public administration in China. First, each component is developed in isolation, without adequate consideration of its relationship to the other components. While the separate components do ultimately sit together, some of the links within the resulting framework are tenuous and lack the formality required to facilitate effective coordination. Second, there is often limited communication between the administrative authorities responsible for related management functions. This phenomenon was present in the water authorities for each case study city, regardless of the presence of an 'umbrella' water affairs bureau—even in such cases there is a lack of internal coordination. For example, the institutional responsibilities for the planned water saving and water use system within the Beijing Water Affairs Bureau are not well defined, with two subdivisions having overlapping roles at times.

These issues indicate the need to consider integrated management approaches when future policy development work is undertaken and to encourage internal dialogue to improve administrative coordination. Leadership on the national stage would be important in this regard, with the recent government restructure, that resulted in more formalized coordination between the Ministry of Water Resources and the Ministry of Environmental Protection on water quality matters, a step in the right direction. Improved institutional coordination could also be served by ensuring there are clear distinctions between those agencies with regulatory responsibilities, such as granting abstraction permits, and those with commercially-driven operational functions, such as the provision of retail water supply services. Other practical strategies to facilitate coordination could include providing guidance on particular administrative matters, such as preparing standard water supply and use contracts to operate throughout the country. This could help coordination by ensuring that there are clear connections between the provisions of the contract and the relevant water abstraction permits and plans.

A more specific issue is that some of the objectives and methods for implementing particular parts of the management framework are not well defined. This has resulted in ambiguities at the local level in terms of what the national requirements are trying to achieve. As a result, some regions simply don't adopt the tools at all, while other regions interpret the objectives in various ways with very different results. For example, while the 2002 Water Law requires the preparation of water supply plans, their scope and objective have not been clearly defined at a national level. Consequently, plans are prepared to comply with the legislative requirements but are not used in practice. The Beijing Waterworks Group Company does not use its water supply plan in the day-to-day operations of the water supply network. In Shanghai, water supply plans are prepared annually by the water utility companies and are used in determining bulk supply agreements with the raw water supply corporations. Similar to Beijing, however, the day-to-day regulation of the water distribution network relies on separate arrangements, with monthly regulation plans used by the Shanghai Water Affairs Bureau's Water Supply Management Department. Water supply plans are not prepared at all in Shaoxing. National guidelines on the role of water supply plans could facilitate more efficient arrangements regarding the day-to-day regulation of water delivery networks. In contrast, the water abstraction permit system is an example of a management tool with a clear objective and mode of operation.

The demarcations between some of the urban water management framework components could also be further clarified, to reduce overlaps, gaps and inconsistencies. For instance, the planned water use and water saving system provides a mechanism to control the use of water by non-residential users of public urban water supply networks. The situation in relation to private water supply systems developed by large users is less clear. The 1994 Urban Water Supply Regulations requires that such users are included in the planned water use and water saving system. At the same time, these users hold a water permit and are therefore required to have an annual water abstraction plan, which caps their annual use. This could also result in conflicting obligations being imposed on the permit holder. A solution could be to amend the 1994 Urban Water Supply Regulations such that the planned water use and water saving system does not apply where the user holds a water abstraction permit.

Similarly, the urban water drainage permit system and the environmental management of discharges to watercourses overlap in the case of wastewater released directly to watercourses in urban settings. Existing regulations are ambiguous in terms of whether authorization is required from the local environmental protection department in addition to an urban drainage permit, which could potentially lead to the imposition of conflicting obligations. If both systems are to apply, their effectiveness will require a high degree of coordination between the relevant agencies to ensure consistency. Alternatively the law could be amended to recognize that environmental aspects of wastewater discharge be managed under only one of the two regulatory mechanisms.

Implementation of National Policies and Regulations

Taking China's urban water management framework as it stands, a number of issues can be identified from the way that, or extent to which, particular systems have been implemented. Of the case studies presented in this paper, Shanghai demonstrates the most progress. The city benefits from a high degree of integration in its urban and rural water management functions, enabling stronger internal coordination between the resources management, urban planning and environmental protection aspects of urban water supply and management. Despite this, there remain particular elements of the framework that could benefit from improved implementation.

The different approaches to implementation can often be attributed to environmental and economic contexts. Beijing has successfully implemented the planned water use and water saving system, whereas its management of water abstraction permits, drainage permits and water supply and use contracts is weak. This is typical in northern China, where the climate and relatively limited water availability has meant that careful use of supplies has been the focus for some time. In contrast, southern cities have greater water resource availability and have tended to be better at adopting water supply and use contracts. This perhaps reflects the financial benefits of ensuring all water use is measured and charged, and the lesser need to restrict consumption levels. The historical rarity of water shortages in the south means that there have been reduced incentives to ensure that abstractors do not take more than is authorized under their permits, and thus the water abstraction permit system, and its link to the water resources allocation planning framework in particular, has not been treated as a priority. The adoption of the water abstraction permit system has also been limited in Beijing, which suggests that the benefits of capping water use through the use of permits and annual abstraction plans may not have been well recognized.

China's approach to ensuring reliable supplies for its urban areas has to date focussed more on constructing new infrastructure than on consideration of the overall water availability and demands at a catchment level. In the cities investigated, and throughout China, water resources allocation planning at both basin and regional levels is incomplete at best (and often entirely absent). Moreover, where such plans do exist, they are not used as the basis for the grant of water abstraction permits as is required under the national-level framework (WET, 2006). Without a formal connection between resources allocation planning and the abstraction permit system, it will become increasingly difficult to ensure urban areas receive the reliability of supplies that is expected: the continued grant of permits, in the absence of a planned limit on total abstractions, will ultimately erode the reliability of existing users.

Again, encouragement from the central government is expected to assist regional administrations appreciate the need to properly manage abstractions under the permit system and, in connection with water resources allocation plans, to ensure that the take of water is both sustainable and appropriate. To support the evaluation of permit applications, urban water use quotas could be revised with a view to establishing a macro-level quota for specific areas (such as industrial zones). This approach could help provide accurate data for setting an appropriate volume of water allowed to be abstracted under a permit. In addition, the importance of managing and supervising water abstraction permits once they have been issued could be given more emphasis. This could include demonstrating the value of preparing annual water abstraction plans in accordance with the approved water quantity on the water abstraction permits.

The system of imposing higher water charges for use that exceeds the authorized volume would benefit from more comprehensive implementation. Further work could be undertaken to identify the appropriate means of collecting such charges from different user groups. For example, users that operate their own water sources and supply networks should have a water abstraction permit and annual water abstraction plan. These plans could contain standard terms requiring the payment of higher water resource fees for abstractions above the authorized volume and, potentially, fines to reflect the importance of the system. For non-residential water users of public urban water supply systems, the existing planned water use and water saving system is probably the most appropriate mechanism. Its full implementation would see inclining block tariffs used on a standard basis in accordance with use allowances under annual water use plans.

Conclusion

Urban water resources management is both highly complex and fundamentally important. It crosses a range of management areas, including water resources allocation planning, urban development planning and environmental protection, each of which has different goals and priorities. A comprehensive urban water management system that can address these different management aspects in a coordinated manner is vital to ensuring that urban communities have access to safe and reliable supplies of water for a range of purposes. This is especially the case in China, where urban development and expansion is continuing at a great pace.

The urban water management framework that has evolved over the past two decades in China has been explored in this paper through three case studies. The analysis has shown that China's management arrangements address the various resource, development and environmental considerations that are necessary in an urban setting and that these components generally fit together to provide a comprehensive management framework. There are particular aspects of this framework, however, that do not integrate particularly well or that have not been prioritized. There are therefore opportunities to enhance China's urban water management, especially in relation to ensuring that urban water use occurs within the resource availability confines determined through the water resources allocation planning and water abstraction permit systems. Seizing these opportunities will require greater levels of communication at both the central and regional levels. It will also need a greater focus on the implementation of the various components of the framework to ensure it operates as intended. Ultimately, commitment on both these fronts will be vital in ensuring that China's cities have the water supplies necessary to allow them to continue to grow and develop.

Acknowledgements

This paper is the result of a project undertaken under the auspices of the Australian Department of the Environment, Water, Heritage and the Arts and the Chinese Ministry of Water Resources, with funding provided by AusAID, the Australian Agency for International Development. The authors of the paper would like to thank the assistance provided by the Beijing Water Affairs Bureau, the Shanghai Water Affairs Bureau, the Zhejiang Provincial Water Resources Department and the Shaoxing Water Affairs Corporation during the case study investigations.

References

Beijing Drainage Group Company (n.d.) *Chief Business*. Available at http://www.bdc.cn/cenweb/portal/media-type/html/user/anon/page/BDC_Chief.page (accessed 21 November 2008).
Beijing Municipal Government (n.d.) *Beijing Figures*. Available at http://www.ebeijing.gov.cn/BeijingInfo/BJInfoTips/BeijingFigures/t965511.htm (accessed 6 October 2008).
Beijing Water Resources Sustainable Use Plan for the Early 21st Century (2001) State Council of the People's Republic of China.
Comprehensive Standard of Sewage Drainage (1996) Ministry of Construction (Standard GB8978).
Contract Law (1999) National People's Congress of the People's Republic of China.
China Daily Online (2003) Shanghai Population Tops 20m, December 5. Available at http://www.chinadaily.com.cn/en/doc/2003-12/05/content_287714.htm (accessed 8 October 2008).
Details on Implementing Water Abstraction Permits (2006) Zhejiang Provincial Government.
Fu, T., Chang, M. & Zhong, L. (2008) *Reform of China's Urban Water Sector* (London: IWA Publishing).
IPART (Independent Pricing and Regulatory Tribunal) (2007) *Literature Review: Underlying Costs and Industry Structures of Metropolitan Water Industries* (Sydney: Independent Pricing and Regulatory Tribunal of New South Wales).
Kessides, I. N. (2004) *Reforming Infrastructure: Privatization, Regulation, and Competition* (Washington, DC: World Bank and Oxford University Press).
Law on the Prevention and Control of Water Pollution (2008) National People's Congress of the People's Republic of China.
Lee, S. (2006) *Water and Development in China: The Political Economy of Shanghai Water Policy* (Singapore: World Scientific Publishing Company).
Liu, W. (2007) Beijing tap water now safe to drink, *China Daily*, July 3, p. 4.

Management Methods on Water Saving (1994) Shanghai Municipal Government.

Ma, X. & Ortolano, L. (2000) *Environmental Regulation in China: Institutions, Enforcement, and Compliance* (Lanham, MD: Rowman & Littlefield).

Methods on Urban Drainage Permit Management (2007) Ministry of Construction.

OECD (Organization for Economic Cooperation and Development) (2000) *Global Trends in Urban Water Supply and Waste Water Financing and Management: Changing Roles for the Public and Private Sectors* (Paris: OECD Publishing).

OECD (Organization for Economic Co-operation and Development) (2007) *OECD Environmental Performance Reviews: China* (Paris: OECD Publishing).

Regulations on Urban Water Saving Management (1988) Ministry of Construction.

Regulations on the Administration of Water Abstraction Licensing and Collection of Water Resources Charges (2006) State Council of the People's Republic of China.

Shanghai Municipal Government (2007) *Basic Facts*. Available at http://www.shanghai.gov.cn/shanghai/node17256/node17432/index.html (accessed 8 October 2008).

Shaoxing Municipal Government (2005) *Shaoxing – China*. Available at http://www.sx.gov.cn/enportal/ (accessed 7 October 2008).

Shen, D. & Liang, R. (2003) *State of China's Water* (Mexico City: Third World Centre for Water Management and The Nippon Foundation). Available at: http://www.thirdworldcentre.org/epubli.html

Shen, D. & Liu, B. (2008) Integrated urban and rural water affairs management reform in China: Affecting factors, *Physics and Chemistry of the Earth*, 33, pp. 364–375.

Shen, D. & Speed, R. (2009) Water resources allocation in the People's Republic of China, *International Journal of Water Resources Development*, 25(2), pp. 209–225.

UN (United Nations) (2008) *The Millennium Development Goals Report 2008* (New York: United Nations). Available at: http://www.un.org/millenniumgoals/pdf/The%20Millennium%20Development%20Goals%20Report%202008.pdf

UNEP (United Nations Environment Programme) (n.d.) *Beijing 2008 Olympic Games: An Environmental Review* (UNEP e-Book). Available at http://www.unep.org/publications/eBooks/beijing-report/Default.aspx (accessed 21 November 2008).

Urban Planning Law (1989) National People's Congress of the People's Republic of China.

Urban Water Supply Quality Regulations (2007) Ministry of Construction.

Urban Water Supply Regulations (1994) State Council of the People's Republic of China.

Water Law (2002) National People's Congress of the People's Republic of China.

Water Resources Fee Collection Management Methods (2007) Zhejiang Provincial Government.

Water Saving Methods (2005) Beijing Municipal Government.

WET (2006) *Water Entitlements and Trading Project (WET Phase 1) Final Report* November 2006 [in English and Chinese] (Beijing: Ministry of Water Resources, People's Republic of China and Canberra: Department of Agriculture, Fisheries and Forestry, Australian Government). Available at: http://www.environment.gov.au/water/action/international/wet1.html

WET (2007) *Water Entitlements and Trading Project (WET Phase 2) Final Report* December 2007 [in English and Chinese] (Beijing: Ministry of Water Resources, People's Republic of China and Canberra: Department of the Environment, Water, Heritage and the Arts, Australian Government). Available at: http://www.environment.gov.au/water/action/international/wet2.html

Wei, D. (2005) Beijing water resources and the South to North Water Diversion Project, *Canadian Journal of Civil Engineering*, 32, pp. 159–163.

Yang, H. & Abbaspour, K. C. (2007) Analysis of wastewater reuse potential in Beijing, *Desalination*, 212, pp. 238–250.

Yongding River Trunk Stream Water Resources Allocation Plan (2007) State Council of the People's Republic of China.

Zhong, L., Mol, A. P. J. & Fu, T. (2008) Public-private partnerships in China's urban water sector, *Environmental Management*, 41, pp. 863–877.

Transferring and Trading Water Rights in the People's Republic of China

ROBERT SPEED

ABSTRACT *Cap and trade systems are becoming increasingly common in water resources management as a mechanism to allow water to move to its highest value use. China too is taking steps down this path with the development of a 'water rights transfer system'. While this system is still in an embryonic stage, a number of government-facilitated projects have already demonstrated its likely form. These projects include the sale of long-term water access rights from one regional government to another, as well as water savings projects in large irrigation districts, with rights to the 'saved' water transferred to industry.*

  These transactions are often occurring in a regulatory environment where water rights and the rules governing them are not clearly defined, and in the absence of a strong framework for managing water transfers. This can create ambiguity over the nature of the right being transferred. There is also significant risk of unintended, adverse impacts—to other water users and the environment—if water is reallocated between users in the absence of defined entitlements to water.

Introduction

Chronic, global water shortages coupled with ever growing human demands for water have highlighted the need to allow for water to be reallocated between users. In the absence of reallocation mechanisms, water can become tied to certain locations or users, and new (water-dependent) development can struggle to access the water it requires in fully allocated systems. Transferring and trading water and water rights is thus seen as an important mechanism to allow for ongoing development and to maximize the value of available water resources. At the same time, reallocating water, especially within heavily developed systems, carries a significant risk of adverse impacts on existing water users, the environment, or both. Management arrangements need to be careful to protect against such outcomes (Xie, 2008; Productivity Commission, 2003).

This article considers the approach being adopted by China's water managers to allow for water to be reallocated amongst different users. The paper initially considers the theory behind water trading and the basic requirements for a water rights transfer system. The paper then describes China's policy approach to this issue and its current water rights and transfer framework. A number of case studies are then described, which represent the country's first efforts at water trading and are a pointer to the future direction of water trading in China. These include transfers at the regional, the abstractor, and farmer (i.e. within irrigation districts) levels. Finally the article concludes with a discussion of the opportunities and challenges for China in implementing a water transfer system.[1]

The Fundamentals of a Water Rights Transfer System

Providing a mechanism for water, and the rights to the water, to be shifted to those regions and sectors where it is most needed has been a major driver to reforms in the water sectors of many countries (Productivity Commission, 2003). In Australia, the United States, Chile and elsewhere, governments have sought to address water-related constraints to development through market mechanisms. In these countries, cap and trade systems have been implemented in an effort to allow water to move to its highest value use (in economic terms) and thus maximize the economic return from the available water resources (*ibid.*). Evidence suggests that there can be significant economic benefits from this approach to water management (Hearne & Easter, 1997; Chong & Sunding, 2006; Young, 2008).

These systems depend on granting water rights and establishing a framework that allows for the trading of those rights. They are designed to encourage individual water users to make their own economically-rational decisions on what to do with their entitlement: whether to use it, to invest in water saving technology (to maximize its value), or to sell it to another user. One alternative to a market-based approach is for water rights to be shifted from one user to another through compulsory reallocation by government. This can be more appropriate where water is required for a public purpose, such as to increase domestic water supplies or water for environmental flows. In such circumstances, the water user whose right is being cancelled or reduced may be entitled to compensation, depending on the relevant government's policy approach and the prevailing legal system.

While water trading as a concept is relatively simple, it relies on sophisticated planning and management systems to support its operation. In most cases, this is in the form of a broader water rights system: that is, a system for clearly identifying the rights of different users to the available water resource, and setting rules to protect those rights from adverse impacts, whether as a consequence of actions of other water users or arbitrary decision-making by government agencies. In simple terms, water rights and trading systems typically consist of:

- *a planning mechanism* which provides the basis for the initial allocation of rights, ideally based on the hydrological and ecological limits of the relevant river or aquifer; and
- *a regulatory system* usually comprised of a licence or permit system (to define individual rights), the conditions attached to them (including seasonal access rules, protections from interference, etc.), and the rules governing trading of the rights (Xie, 2008).

Where market-based trading has been implemented, trading rules typically govern the extent to which water rights can be transferred between locations, sectors and priority groups, with rules designed to protect against adverse impacts on either the environment or third parties (Xie, 2008). There is the potential for trading of both long-term rights to water as well as the annual water available to a right holder. In the latter case—often referred to as temporary trading—the seller trades only a volume of water for the current water year, while retaining the long-term right (and hence will be entitled to water again in following years). In Australia, temporary trading accounts for around 90% of all water trades (Young, 2008).

China's Water Allocation and Transfer Framework

This section considers China's water transfer system in light of (1) the rights to water held by different governments, individuals and entities, and (2) the scope for transferring those rights. The framework for the allocation and management of China's water resources is established by the 2002 Water Law. The water allocation system is described in detail by Shen and Speed (2009) but in summary, the framework can be considered in terms of:

- *Regional water rights*—the right to a share of a common water resource, granted to an administrative region under a water allocation plan. These are in effect the right of an administrative region to allocate its share of the common resource amongst water abstractors or to sub-regions (2002 Water Law, Articles 44–47);
- *Abstractor rights*—granted via a water abstraction permit system to entities (including factories, water supply companies and irrigation districts) as the right to take water from a river or groundwater system (2002 Water Law, article 48); and
- *User-level rights*—granted, for example, to farmers within an irrigation district to define their share of the district's water allocation (Code of Practice for Technical Management of Irrigation and Drainage Engineering, 1999).

In practice, rights at these three levels have been granted to varying degrees and in different ways across China. In some regions sophisticated allocation systems have been implemented, while in others management is at a more basic level (Shen & Speed, 2009). Where water rights have been established, water transfers are theoretically possible between each of these three levels. That is, water can be transferred between regions, between abstractors and between farmers. To date in China, transfers have happened to a limited degree, and have been more the result of ad hoc initiatives rather a systematic attempt to introduce a water rights transfer system, as can be seen from the examples described later in this paper.

China's Policies and Laws Relating to Water Transfers

China's 11th Five-Year Plan (2006–10) places a priority on improved water resources management. It specifically requires the establishment of "an initial water right distribution system and a water right transfer system" (State Council, 2006). Both prior to and since the plan's introduction in 2006, there have been efforts within China's Ministry of Water Resources to move towards a rights-based water management system (Wouters *et al.*, 2004). At the same time there have been a number of Ministerial guidelines and directives issued on water transfers, both region-specific and general.[2] However, despite this, for the most part existing laws provide little direction in respect of the requirements for transferring water rights.

Regional water rights are granted via water resource allocation plans and as such these plans are the definitive record of regional rights to water: that is, regions are not issued separate documentation identifying their entitlement. Consequently, transfers (theoretically) require an amendment to the water allocation plan to vary the shares of the relevant regions. Any amendment will require the same process that was followed in making the initial plan and thus requires the approval of the People's Government at the appropriate administrative level.

As such, a plan could be amended either at the behest of a superior level of government, to reallocate water between its regions in accordance with new regional priorities, or as a result of an agreement between two regions. For example, the water allocation plan for the Jin River (Jinjiang) in Fujian Province is currently under review. As part of this process, those counties allocated 'surplus' water under the previous plan may be able to negotiate to sell some of that spare capacity to other counties (Water Entitlements and Trading Project (WET), 2006). As an alternative to this somewhat cumbersome approach—particularly where a plan is not due for revision—regional governments have sought to circumvent the plan requirements through contractual arrangements, as described later in the case of the Dongyang-Yiwu water transfer.

At the abstractor level, entitlements to water are managed through an abstraction permit system. Article 27 of the 2006 Decree on the Administration of Water Abstraction Licensing and Collection of Water Resources Charges (or the Water Permit Regulation) specifies the circumstances in which a water permit (or part thereof) can be transferred. It provides that where a permit holder reduces their water use—through for example "adjustments to production or industrial structure, innovation in technologies or the promotion of water-saving measures" —the holder can transfer the saved water "with compensation", i.e. with some payment to the transferor.

This is the only kind of trade that is specifically allowed at the abstractor level. That is, a permit holder is not allowed, as of right, to transfer a water permit simply because the abstractor no longer wants or needs the water—it is only where actual water use has been demonstrably reduced that part of the water rights can be transferred. To date, the best examples of these kinds of transfers have involved water savings through the lining of irrigation channels, with the water saved as a result of reduced transmission losses subsequently transferred to industrial users.

Significantly, earlier drafts of the 2002 Water Law included further provisions in respect of the transfer of water rights. However, the proposal to include these provisions generated significant controversy and the provisions were not included in the law as made (Wouters *et al.*, 2004). At the irrigation district level, rights have only been granted at the farmer level in a few select pilot sites. In these areas, local administrative arrangements generally allow for farmers to transfer their annual water tickets to other farmers, however, there is no firm legal basis or framework for these transactions (WET, 2006). This type of allocation and transfer system is discussed further below.

The following sections describe in more detail examples of different water transfers. These examples have been test cases: in some instances, they have been facilitated by the central government; in other cases, they have been achieved by local governments within the current management framework, through innovation and to meet particular local demands. Regardless, they show the direction China is taking as it moves to allow for water transfers. They also highlight some of the challenges that will need to be addressed as China's water rights transfer system develops.

Pilot Water Transfer Projects

The Transfer of Regional Water Rights

The agreement reached in 2000 between Dongyang and Yiwu Counties is widely regarded as China's first example of a regional water transfer (Gao, 2006). The counties are both located in the Jinhua Prefecture in Zhejiang Province in China's wet southeast. Contamination from pollution meant that Yiwu's existing water sources were struggling to meet its water requirements, particularly in light of growing demand associated with economic development. Neighbouring Dongyang was seen as a potential source of clean water.

In December 2000, the regional governments of Dongyang and Yiwu signed a contract, whereby Dongyang agreed to supply Yiwu with 50 million m^3 of water per year from the Hengjin Reservoir in Dongyang. A pipeline to provide the water to water users in Yiwu was completed in 2005. The pipeline was constructed and paid for by Yiwu, although Dongyang retained operational responsibility for the part of the pipeline within its regional boundaries. The contract provided for "the permanent transfer of the water use right" for 50 million m^3, in return for a lump sum payment to Dongyang of RMB200 million. Dongyang retained ownership of the reservoir as well as operational and maintenance responsibility, with an operation and management fee of RMB0.1 per m^3 payable by Yiwu, calculated based on the actual volume of water supplied each year. The contract also requires Dongyang to guarantee the quality of water supplied. The arrangements were formalized through a water abstraction permit, granted to Yiwu County allowing it to take water from the Hengjin Reservoir[3] (Gao, 2006; WET, 2006).

Several other similar regional transfers, via contract, have also occurred in Zhejiang Province. Cixi County has signed separate agreements with both Yuyao and Shaoxin Counties. Both of these are contracts to supply water to Cixi, where pollution has reduced the level of available water resources. A notable difference though is that Cixi plans to meet its long-term water supply needs by improving water quality through better environmental management. As such, the contracts have a limited life and there is no permanent transfer of water rights. In the case of Yuyao, the contract is for 15 years, which reflects the expected lifespan of the supply infrastructure (Gao, 2006; WET, 2006).

Saving and Transferring Water in the Yellow River Basin

The Yellow River, as the dominant water source for the populous but water-scarce north, has been at the forefront of innovations in China's water allocation, water regulation and, more recently, water transfer systems. The Yellow River Water Allocation Plan, approved by the State Council in 1987, was the first of its kind in China. The plan allocates an average annual volume of 37 billion m^3 amongst the ten provinces that use water from the basin, and in the process caps the total consumption from the trunk stream of the river (Shen & Speed, 2009). The booming economy of northern China has seen demand increase and water become a limiting factor to development in many parts of the basin. In the Hui Autonomous Region of Ningxia, the 4 billion m^3 allocated to the region has been fully allocated to different users (via abstraction permits), more than 90% of which is for agricultural purposes. This has limited the development of new, water-dependent industries. Similar issues exist in neighbouring Inner Mongolia.

To address these problems, the government—at both the central and local level—has facilitated a series of water-savings initiatives within large irrigation districts. This has involved lining hundreds of kilometres of earthen channel to reduce transmission losses. This 'saved' water has then been transferred to industrial users, with the industries in turn paying the cost of the channel lining and ongoing maintenance. In Inner Mongolia, the first transfers were conducted in 1998, involving the sale of water saved in two different irrigation districts to two power stations. Since then, a series of guidelines have been issued, by both central and local agencies, to provide a more structured approach to the transfer process.[4]

Government agencies have been central to the process which, in broad terms has consisted of:

- the local water agency undertaking an assessment of priorities for channel lining, the potential volume of water that could be saved, and the implementation cost;
- a public invitation for enterprises in need of water to submit proposals identifying their water requirements;
- selection by the government of those enterprises that would be granted permits to the saved water (once channel lining was completed). The successful applicants were chosen based on whole-of-government development priorities. Notably, applicants did not submit a price they were prepared to pay—they were instead advised what the cost would be. The water on offer was oversubscribed, meaning only 30 out of more than 50 applicants were successful; and
- the successful applicants signing contracts with the prefecture government, agreeing to fund the channel lining in return for being granted water rights (Gao, 2006).

Initially, contracts between the industries and the local government specified which sections of channel in an irrigation district the particular purchaser was funding. Work on that section would then only take place once payment was forthcoming. However, the process has subsequently been streamlined, and there is no longer a direct link made between an individual purchaser and a particular section of channel.

Between April 2003 and July 2006, 30 potential purchasers of water rights were identified in Inner Mongolia, with contracts signed by 16 of them, involving a total funds transfer of RMB840 million for water rights totalling 153 million m^3 (WET, 2006). The purchasers pay a price that covers not only the cost of channel lining, but also the ongoing operations and maintenance. In some instances these operations and maintenance costs are to be paid annually by the purchaser, meaning that there is an ongoing contractual relationship between the irrigation district and the purchaser.

Once channel lining is completed (with the work generally arranged by the local government), the work is assessed by the Yellow River Conservancy Commission (YRCC) to confirm that the 'savings' have been realized. The YRCC is then responsible for adjusting the abstraction permit for the irrigation district (i.e. to reduce it by the saved/transferred volume) and grant a new abstraction permit to the purchaser. Through this process, in Hangjin Irrigation District in Inner Mongolia, the district's allocation (and abstraction permit) has been reduced from 410 million m^3 to 280 million m^3.

A similar approach has also been adopted in Ningxia, where water transfers have been undertaken in accordance with the Ningxia Yellow River Water Rights Conversion Master Plan. According to the plan, by 2010 an additional 330 million m^3 will be made available to industry by channel lining of irrigation districts in Ningxia. This will increase to 494 million m^3 by 2015 (representing more than 10% of Ningxia's total allocation from the Yellow River). The transfer process is broadly the same as that described above for Inner Mongolia. The plan also states a series of fundamental principles to be applied in implementing the water transfers:

- "gross volume control"—the transfer must not result in an increase in consumption above the 4 billion m^3 allocated to the region from the Yellow River;
- "clarified water rights"—the seller must have a water abstraction permit;
- "unified management"—the Ministry of Water Resources, the YRCC and the Ningxia Water Resources Department must agree on management of the transfer;
- "consultation with all parties democratically, openly, equally and fairly";
- "transfer with financial compensation";
- "combining government regulation with market mechanisms" (Ningxia Yellow River Water Rights Conversion Master Plan).

These principles, together with the methodology described above, are indicative of the general approach being adopted to water transfers in China.

Water Transfers within Irrigation Districts

The grant of water rights at the farmer level has only occurred to a limited extent to date, principally as part of pilot water efficiency programmes in some of the most water-scarce regions. The Liyuan Irrigation District in the Hei River Basin in Gansu Province is a typical example. Within the district, 'water certificates' are issued to individual households, showing the allocation assigned to the household and the conditions for use. On an annual basis, allocations are assigned using a 'water ticket' system. Individual farmers purchase water tickets from the irrigation district management agency (or the local water user association) for each irrigation cycle. The volume they can purchase is limited by the volume on their water certificate and the annual volume available to the district that year. The water tickets are then submitted to the local irrigation district officials prior to water being delivered, via the channel system, to their farm. They act as both an ordering system and a mechanism for prepayment of water charges (WET, 2006).

Water tickets may be traded freely and easily as they are not user- or location-specific: farmers can sell their water tickets (but not water certificates) to other farmers in the district and no administrative approval is required. The ticket can then be passed to the district management agency in return for the supply of water. In practice, there have been few instances of trading of water tickets, and anecdotal evidence suggests that water tickets are sold between neighbours, adjusting where one has over-ordered. Primarily this is because the water available to farmers in these districts is normally sufficient to meet their requirements without the need to purchase additional water (Gao, 2006). This approach suggests the likely future of farmer-level water rights and water transfers.

Opportunities and Challenges for China

The water transfers made to date in China have been successful to the extent that they have allowed water and water rights to move to where they are most needed. The transfers have been born out of necessity, and have been crafted to meet the particular needs of individual situations. As such, while these transactions may have achieved their primary objective, the implementation of a transfer system with a broader application will need to consider the long-term consequences of these types of dealings.

The Importance of Well-defined Water Rights

As noted earlier, the establishment of water rights—clearly defined and protected entitlements to a share of the available water resource—is vital to the operation of a water market (Productivity Commission, 2003). However, they are also important to support government facilitated (i.e. non-market) transfers. In either case, wherever entitlements to water are being reallocated, water rights are critical:

- to define what is being transferred;
- to provide confidence to the holder of a right that they will continue to be able to access their right (i.e. take water), particularly where a payment is being made to purchase the right; and
- to support the water planning system, by ensuring that the total rights granted are within the sustainable hydrological and ecological limits of the system.

Despite efforts since the introduction of the 2002 Water Law, water rights in most cases in China are not well established, either at the regional, abstractor, or farmer level. Rights have often not been granted at all, and where they have, the rules surrounding the rights are often ambiguous (WET, 2006).

As such, where transfers have been made, it is not always clear what has been transferred and hence what are the rights of the purchasing party. The responsible government agencies have attempted to overcome this issue through a variety of ad hoc measures, guidelines and contracts. This approach has the potential to result in problems down the line and to limit the applicability of these methods to other regions in China. Some specific issues associated with the pilot water transfers are discussed further below. Fundamentally though, these all relate to the importance of clearly defined rights and the systems that govern them.

The Importance of Defining Water Sharing Rules

In the case of the regional water transfer between Dongyang-Yiwu Counties, there are no clear rules for determining how the water available from Hengjin Reservoir in a given year is to be shared, whether between the regions or other permit holders, nor is there an allocation plan which limits the volume of water that may granted (under permit) for abstraction from the reservoir. At the same time, there is no specified level of reliability for the water right purchased by Yiwu. Therefore, there is potential for future disagreement during periods of water shortage or where further abstraction permits are granted.

These issues exist because the sale was, quite explicitly, of the long-term water right in respect of 50 million m^3 of water, while for the most part the contract between the counties is more in the form of a water supply agreement. The distinction is important: under a supply agreement, Yiwu County would simply be entitled to receive its annual supply of water (with whatever financial penalties may apply for a breach of the agreement). However, with the purchase of the right itself, Yiwu gained a right not to a volume of water but, more accurately, to a share of whatever is available from Hengjin Reservoir. In granting water rights, it is necessary to identify not only the volume but all the other rules that affect that right, and other water rights from the same source. These rules will determine actual availability on an annual basis. Anecdotal evidence suggests that disputes are already arising over sharing the available water from the reservoir (WET, 2007).

The issues discussed above are compounded by the lack of clear separation of institutional responsibilities with respect to water resources management and water supply. Where regional water transfers have occurred, the contracting parties in some cases have been the local governments (or their water management agencies), while in others it has been government-owned water supply companies. As a consequence, and in the absence of clear operational and water sharing rules, these rules have in some cases been incorporated within contracts for the sale of water entitlements, to provide a level of certainty to the purchaser. This has resulted in rules affecting the availability of water in a region—rules in respect of resource management issues—being included in a commercial contract. This raises issues for third parties as to how their water rights might be affected by the water sharing rules included in these contracts.

In a similar vein, contracts for the sale of water entitlements have bundled supply arrangements and the maintenance of infrastructure along with the entitlement to water. Any future dealings with these entitlements—for example, if Yiwu identified a better water source and wished to sell its rights to another party—would require the unbundling of the right to water from the contractual obligations.

Third Party Impacts of Water Transfers

Agriculture is by far the biggest user of water in northern China, accounting for 80–90% of water use in some of the regions along the Yellow River. With the efficiency of many irrigation districts at less than 35%, these are obvious targets for finding additional water to meet domestic and industrial needs. Doing so through lining channels, or other efficiency measures which will allow the same net volume of water to be available for irrigation, is a logical step. There are, however, several key issues that will need to be addressed to ensure this occurs in a sustainable and equitable manner.

Third party impacts, including environmental impacts, need to be better considered as part of any transfer process. Where water saved through channel lining has been transferred, little consideration has been paid to the interrelationship between surface and groundwater. It is possible that water that was being 'lost' through channel seepage was in fact recharging local groundwater or ultimately returning to form part of the base flow of the Yellow River (Xie, 2008). In any transfer system, the impacts on the groundwater system and those communities and ecosystems that depend on it should be properly assessed as part of the transfer process (Dinar *et al.*, 1997; Perry *et al.*, 2007). Likewise, the overall affect on flows in the Yellow River itself need to be assessed. Doing so will require a water allocation plan, and transfer rules, that recognizes and regulates water from all sources within the basin. Failure to do so could result in either a net increase in water taken from the river, a fall in groundwater levels, or both.

As part of the water savings exercise in Hangjin Irrigation District, some of the water saved was set aside for ecological purposes, to provide for groundwater-dependent vegetation that is likely to suffer from falls in groundwater levels. This demonstrates an awareness of some of the possible consequences of this initiative, although it is not clear that the reserved water will be adequate to meet environmental and domestic requirements if there are significant falls in water levels (WET, 2007).

Expanding Water Markets

Water transfers at the abstractor level are currently limited to 'saved' water. From a policy perspective, this restriction can been seen as necessary for three reasons:

- to prevent a reduction in agricultural production, as a result of farmers cashing in their water rights by selling them to water-hungry industries;
- to prevent water permit holders from reaping a windfall gain, where permits have been granted for a volume in excess of that actually required or used by the permit holder;[5] and
- to prevent an increase in overall water use: this may occur if (again) permits have been granted for volumes in excess of actual water usage. Trading the unused portion of the permit could increase the actual water taken from the system, with consequences for the environment or the reliability of supply to other users.

There are, of course, alternative policy mechanisms to address the above concerns. First, these could involve establishing water allocation plans (and caps), and by reviewing and if necessary revising the volumes granted under existing permits. Once these are in place, water managers could allow for trading to occur with fewer restrictions, while having confidence that there will not be a growth in overall water consumption. This could then allow, for example, water to be transferred where a water user changes the purpose for which it is used (and not just the production process). This in turn could help drive structural adjustment, i.e. changes to the sectors that use water, rather than just improving efficiencies within sectors.

Where there remain concerns over the impacts on particular sectors, such as the risk of lost agricultural production, specific trading rules could be implemented to prevent such outcomes. This could be via sector quotas, to limit the volume of water that can be transferred by or to one sector. This type of approach has been applied in water legislation elsewhere to protect sectoral interests.[6]

Farmer-level Transfers

Water rights management, and the trading of rights, presents special challenges in China due to the huge number of small 'farms' and individual water users: Hangjin Irrigation District, a typical, medium-sized district for northern China, covers 23 000 hectares and is home to around 30 000 farmers. In southern China farms are often much smaller. Allocating and managing water rights at the farmer level is thus a significant administrative challenge.

Similarly, without an appropriate system for facilitating farmer-level trades, transaction costs are likely to be prohibitively high, given the small volumes of water that will be traded. This applies particularly in the case of trading by farmers to users outside the district. The introduction of some form of water bank may provide solutions to this issue (Frederick, 1993). Such a mechanism could allow for farmers to pool any water efficiency gains and sell that spare capacity to others water users (WET, 2007).

Temporary Trading of Water Rights

The pilot projects to date have focused on the transfer of long-term rights to water.[7] Experience in Australia suggests that some of the greatest benefits from trading water rights can come from the temporary water market—trading of annual allocations of water, rather than the long-term right. This market accounts for around two-thirds of all water trading in Australia (NWC, 2008). Future expansions of China's water transfer system might benefit from allowing greater flexibility for water users to transfer their entitlements to water on an annual basis.

Conclusions

As China's water managers try to keep up with the demands of the country's growing economy and population, the pressure to provide for water transfers will continue to rise. China should take care to ensure that a comprehensive and robust system for allocating and managing water rights is in place before progressing too far down the water transfer path. Any transfer system will also need to provide adequate protection against adverse social consequences and impacts on other water users or the environment.

While the application of market forces has proved successful in other countries, it may not be the way forward for China, at least for the time being. Past examples of government-facilitated water transfers—water trading 'with Chinese characteristics'— appear more in keeping with the country's culture and political system, and have been successful in achieving their objectives. These pilot cases have allowed the water to shift to those users who need it, and required them to pay for the privilege, while at the same time allowing the government to maintain a high level of control over the process and ensure that the reallocations meet broader strategic goals. Over time, the introduction of a more comprehensive water rights system is likely to remove the risks associated with water trading, allow the government to step back into a regulatory role, and provide greater freedoms and incentives for water to move between uses and users.

Acknowledgements

This paper is the result of a project undertaken under the auspices of the Australian Department of the Environment, Water, Heritage and the Arts and the Chinese Ministry of Water Resources, with funding provided by AusAID, the Australian Agency for International Development.

Notes

1. In this article, the phrase 'water transfer' is generally used, in preference to 'water trade', when referring to the reallocation of water rights in China, in recognition of the role of the government—as distinct from a free-market approach—in water reallocation.
2. For example, Guidance for the Implementation of Water Rights Transfer of Yellow River in Hui Autonomous Region of Ningxia, issued by the Ministry of Water Resources; Guidelines for the Water Rights Transfer of Main River of Yellow River in Hui Autonomous Region of Ningxia and Inner Mongolia Autonomous Region, issued by the Ministry of Water Resources; and Notes of Implementation Method for Yellow River Water Rights Transfer (trial).
3. The permit was granted by the provincial water resources department, in accordance with the Water Permit Regulation, which requires that permits for volumes of this size be granted at the provincial level. Normally a provincial water resources department would not grant a permit allowing a regional government to take water

from a reservoir belonging to another region, but the contract was sufficient basis to convince the provincial department to do so.

4. Guidelines for the Water Rights Transfer of Main River of Yellow River in Hui Autonomous Region of Ningxia and Inner Mongolia Autonomous Region; Notes of Implementation Method for Yellow River Water Rights Transfer (trial).

5. Permits may have been granted in the past for greater volumes than was required due to limited scrutiny. The consequences were minimal as the permit holder could not physically take more than necessary (due to other restrictions, such as the approved abstraction works) and could not legally transfer the surplus volume.

6. For example, the Queensland Water Act 2000 allows water trading rules to include limits on the volume of water that can be transferred between different uses. The provision was included specifically to meet the agricultural sector's concerns of the social consequences from water being transferred from agriculture to industry.

7. Except in the case of farmer-level trading of water tickets. However, as noted, in practice there have been few if any actual trades of this kind.

References

Chong, H. & Sunding, D. (2006) Water markets and water trading, *Annual Review of Environmental Resources*, 31, pp. 239–264.

Decree on Yellow River Water Resources Regulation, State Council of the People's Republic of China. Available at: http://www.yellowriver.gov.cn/ziliao/zcfg/fagui/200612/t20061222_8784.htm

Dinar, A., Rosegrant, M. W. & Meinzen-Dick, R. (1997) *Water Allocation Mechanisms: Principles and Examples* World Bank Policy Research Working Paper No. 1779 (Washington, DC: World Bank).

Frederick, K. D. (1993) *Balancing Water Demands with Supplies: The Role of Management in a World of Increasing Scarcity* World Bank Technical Paper No. 189 (Washington, DC: World Bank).

Gao, E. (2006) *Water Rights System Development in China* [in Chinese] (Beijing: China Water and Hydropower Publishing).

Guidelines for the Water Rights Transfer of Main River of Yellow River in Hui Autonomous Region of Ningxia and Inner Mongolia Autonomous Region, Ministry of Water Resources (2004) No. 159.

Guidelines on Technical Management of Irrigation and Drainage Project (1999) Available at http://www.cws.net.cn/law/guifan/SL246-1999/ (Accessed 20 September 2008).

Hearne, R. & Easter, K. W. (1997) The economic and financial gains from water markets in Chile, *Agricultural Economics*, 15, pp. 187–199.

Ningxia Yellow River Water Rights Conversion Master Plan (2005) Guidance for the Implementation of Water Rights Transfer of Yellow River in Hui Autonomous Region of Ningxia. Issued by the Ministry of Water Resources. Water Resources Document no. 159, May 18th, 2004. Available at: http://www.hwcc.com.cn/newsdisplay/newsdisplay.asp?Id=101638 (accessed 19 May 2009).

Notes of Implementation Method for Yellow River Water Rights Transfer (trial), Yellow River Water Resources Commission (2004) No. 18.

NWC (National Water Commission) (2008) *Australian Water Markets Report 2007–2008* (NWC, Canberra). Available at: http://www.nwc.gov.au/resources/documents/AWMR2007-08COMPLETE.pdf (accessed 19 May 2009).

Perry, C. J., Rock, M. & Seckler, D. (1997) *Water as an Economic Good: A Solution or a Problem?* Research Report 14, Colombo, Sri Lanka: International Irrigation Management Institute (IWMI).

Productivity Commission (2003) Water rights arrangements in Australia and overseas. Commission Research Paper, Melbourne: Productivity Commission.

Regulations on the Administration of Water Abstraction Licensing and Collection of Water Resources Charges (2006) State Council of the People's Republic of China. Available at: http://www.mwr.gov.cn/zcfg/xzfg/20060221000000967778.aspx (accessed 30 November 2008).

Shen, D. & Speed, R. (2009) Water resources allocation in the People's Republic of China, *International Journal of Water Resources Development*, 25(2), pp. 209–225.

State Council (2006) *National Economic and Social Development Plan for Eleventh-Five Year Period, 2006–10*. Available at http://english.gov.cn/special/115y_fd.htm (accessed 15 November 2008).

Water Law of the People's Republic of China (2002) National People's Congress of the People's Republic of China. Available at: http://english.gov.cn/laws/2005-10/09/content_75313.htm (accessed 20 November 2008).

WET (2006) *Water Entitlements and Trading Project (WET Phase 1) Final Report* November 2006 [in English and Chinese] (Beijing: Ministry of Water Resources, People's Republic of China and Canberra: Department of Agriculture, Fisheries and Forestry, Australian Government). Available at: http://www.environment. gov.au/water/action/international/wet1.html.

WET (2007) *Water Entitlements and Trading Project (WET Phase 2) Final Report* December 2007 [in English and Chinese] (Beijing: Ministry of Water Resources, People's Republic of China and Canberra: Department of the Environment, Water, Heritage and the Arts, Australian Government). Available at: http://www.environment.gov.au/water/action/international/wet2.html.

Wouters, P., Hu, D., Zhang, J., Tarlock, D. & Andrews-Speed, P. (2004) The new development of water law in China, *University of Denver Water Law Review*, 7(2), Spring 2004, pp. 243–308.

Xie, J. (2008) *Addressing China's Water Scarcity: A Synthesis of Recommendations for Selected Water Resource Management Issues* (Washington, DC: World Bank Publications).

Approaches to Providing and Managing Environmental Flows in China

XIQIN WANG, YUAN ZHANG & CASSANDRA JAMES

ABSTRACT   *This paper reviews the course of research on environmental flows in China. It briefly summarizes the history of environmental flows research and introduces twenty approaches used in China to calculate environmental flows. This includes methods adapted from overseas applications and those developed in China to tackle specific environmental issues. The paper gives examples of the implementation of environmental flows in China and identifies some of the deficiencies in environmental flow methodologies. Finally, it discusses obstacles facing the successful implementation of environmental flows in China and suggests the steps required to relieve these impediments.*

Introduction

River systems are some of the most significant ecosystems in nature, possessing diverse values and service functions. The continuing existence of these values and functions depends upon maintaining a healthy river ecosystem. Driven by social and economic benefits, however, humans have influenced, occupied and controlled water resources excessively in recent decades.

In 1949, China's water requirements were 103.1 billion m^3, which increased to 204.8 billion m^3 over the next 10 years. More recently, between 1980 and 1997, water requirements increased by approximately 30%, resulting in an average yearly increase in abstraction of 6.64 billion m^3 (Liu *et al.*, 2001). According to statistics provided by China's Commission on Dams (Jia *et al.*, 2003), by the end of 2003 there were 4694 dams in existence or under construction, with a wall height greater than 30m and a capacity of about 584.3 billion m^3. This represents 20% of the total runoff volume of all Chinese rivers. In addition to the accelerated exploitation of large rivers like the Yangtze and Yellow Rivers, China has also developed water resources in many smaller river systems and had over 20 000 large dams (i.e. with wall heights greater than 15m or volumes of more than 3 million m^3) by the end of the 20th century, which make up 44% of the world total (WCD, 2000).

Concern for the health of rivers in China has heightened with some high profile incidents in recent decades, including the drying up of the mouth of the Yellow River. Between 1986 and

2006, water volumes abstracted for human consumption from the Yellow River were maintained continuously at about 30 billion m^3—about 60% of the natural runoff volume of the Yellow River (45.0 billion m^3) whilst the amount of incoming pollutants to the Yellow River doubled. The reduced water volumes within the Yellow River have greatly weakened the river's ability to flush sediment through and dilute pollutants. This has not only jeopardized the Yellow River's ecological health but also seriously restricted sustainable development of the regional economy (Liu et al., 2008). In another example, the water volumes in the Tarim River in northwest China have decreased from 1.353 billion m^3 in the 1960s to 0.267 billion m^3 in the 1990s, i.e. a reduction of 80% over 40 years (Chen et al., 2008). At the same time, groundwater levels in the Tarim Basin have fallen substantially and the desert vegetation has become severely degraded, critically injuring its biological diversity (Chen et al., 2008).

Water resource development (WRD) in China has impacted on river systems in a multitude of ways. River flow volumes have been reduced whilst seasonal patterns of flow, rates and durations of rise and fall, water velocities and patterns of flow variability have altered. Changes in flow are compounded by issues related to WRD such as the impacts of construction of barriers (i.e. dams, reservoirs and weirs) on movement and migration, the fragmentation of river systems, and alterations to water temperatures, water quality, sediment transport and aquatic habitats (Liu & Yu, 1992; Zhong & Power, 1996). Such factors have influenced the abundance, composition and dominance of many riverine species (e.g. Zhong & Power, 1996; Fu et al., 2008a). In addition, many Chinese rivers face severe threats from factors such as habitat loss and/or degradation, desertification, over-exploitation of river resources, invasive species, climate change and pollution. Recent water quality monitoring (reported in China's State of the Environment Report, 2006 (Ministry of Environmental Protection of the People's Republic of China, 2007)) revealed that amongst the 741 surface water sections monitored by the national environmental monitoring network, 40% were graded I–III, 32% were IV–V grade, whilst 28% failed to meet grade V. (According to the national surface water quality standards of China [GB3838-2002], grades I-III are suitable for drinking, grade IV for industrial and recreational use and grade V is suitable only for agricultural use).

Improving environmental flows is recognized as a critical step in order to sustain natural river ecosystems, protect biological diversity and to restore rivers which have been damaged by excessive water abstraction and inappropriate management. Consequently, environmental flows have become a key focus of research. Environmental flows studies in China have been carried out since the 1980s, making significant progress through nearly 30 years of research. This paper summarizes developments in environmental flows research and its application to date, and analyses some of the critical factors preventing the implementation of environmental flows. Finally, the paper suggests some of the future steps, including research needs, required to relieve these impediments.

A Brief History of Environmental Flows in China

Research on environmental flows in China dates back to the 1970s, when water pollution problems were already evident. Early research primarily focused on requirements for improving water quality. For instance, the unpublished report "Preliminary Exploration of Environmental Water Requirement", produced by the Yangtze River Water Resources Protection Science Research Institute, considered the water quantities necessary to control (i.e. dilute) water pollution.

During the 1980s China experienced a period of rapid population growth and economic development. Over the decade, about 90 large-scale reservoirs were built to satisfy the increasing requirements for irrigation and urban water supply. Environmental problems in rivers and their associated floodplains and wetlands caused by WRD grew during this period, and thus the concept of environmental hydro-engineering was proposed. This concept emphasized the need to take environmental impacts into account during hydraulic engineering projects (Shen, 1988). At the same time, increasing attention was being paid towards coordinating the exploitation, utilization, conservation and administration of water resources. Investigations were begun on how to improve the downstream river-course environment by manipulating reservoir control patterns (Fang, 1988). These developments established a favourable foundation for later environmental flows research.

In the 1990s, faced with severe environmental problems—including general river ecosystem degradation, the contraction and disappearance of lakes and wetlands, the drying up at the mouth of the Yellow River and the loss of floodplain forests along the Tarim River—the Chinese Ministry of Water Resources recognized the need to consider environmental water requirements in the allocation of water resources. In the industrial standard "Environmental Impact Assessment of River Basin Planning" [SL45-92], produced by the Ministry of Water Resources in 1993, environmental water requirements were formally included as a compulsory consideration during water resource planning for environmentally vulnerable areas (Recycled Water of Huai River Engineering Ltd. and Planning and Design Institute, 2006). Government departments could now attach importance to environmental flows.

Meanwhile, research widened and progress was made on refining environmental flows concepts and calculation methods (Li & Zheng, 2000; Wang et al., 2001a; Yan et al., 2001). The publication of research findings from the State Key Sci-Tech Programmes in China's Ninth Five-Year Plan (Liu, 2000) and the "Strategic Report on Sustainable Development of China Water Resources" (Qian & Zhang, 2001) together probably represent the point at which research on environmental flows in China was properly initiated. At the same time, the Water Law, newly revised in 2002, emphasized the need to consider environmental water requirements as part of the allocation and planning process (2002 Water Law, Articles 21 and 22).

From the beginning of the 21st century, with theories and concepts of environmental flows maturing rapidly, research turned to combining theories and practices whilst the direction and focus of studies also began to shift. Scholars started to reassess environmental water requirements combining both water quantity and water quality requirements (Wang et al., 2006), investigating relationships between aquatic species' life histories and flows (Xu et al., 2005; Zhang et al., 2007; Fu et al., 2008b), and studying water allocation in terms of both environmental and socio-economic water requirements (Wang et al., 2007a; Hao & Shang, 2008). Assessments moved from being purely hydrological methods to biological habitat methods, with environmental flow recommendations now more closely linked with the requirements of aquatic species.

Scholars and water managers began investigating how to combine the theory of environmental flows with practice and to discuss the application issues with water resources managers (Water Entitlements and Trading Project (WET), 2007). This resulted in a number of publications and reports covering these issues (Liu, 2000; Yang et al., 2003; Liu, C.M. et al., 2006; Wang, 2007). Alongside the continuing progress made on environmental flow theories and methods, environmental flows have been incorporated into management arrangements in several basins, with recognized ecological benefits

(Feng *et al.*, 2007; Qiao *et al.*, 2007; Ge *et al.*, 2008; Dispatch Act of Water Quantity from the Yellow River, 2006).

A Review of Environmental Flow Methodologies Applied in China

Environmental flow methods in use in China include the direct application of methods developed overseas (for example the widespread application of the Tennant method for water resource planning), modification of existing methods (usually adapted from overseas applications) and methods developed in China to tackle specific environmental issues. There are approximately 20 different calculation methods that have been used in China to date (see Appendix: Table 1), of which the Tennant method (Tennant, 1976) is one of the most heavily utilized, due to its simple methodology and its ease of application.

A large number of the methodologies in use have been modified from overseas methods (Tharme, 2003) requiring some tailoring to the specific Chinese situation. Methods in this group include the Monthly (yearly) Guarantee Rate method (Wang *et al.*, 2003), the Average Flow in the Lowest Flow Seasons method (Water Resources Protection Bureau of Haihe River Basin, 2000) and the Ecological Hydraulic Radius method (Liu, C.M. *et al.*, 2007). Other methods have been developed to deal with specific situations such as sediment transport (e.g. the Minimum Sediment Transmission Volume in Flood Seasons method and the Experience method) or water quality and pollution issues (e.g. the Water Quality Model method, the Environment Function method and the Minimum Flow in Dry Season method) (see Table 1).

Environmental flow methodologies can broadly be grouped into six types: hydrological methodologies, hydraulic methodologies, habitat simulation methodologies, sediment transport methodologies, holistic methodologies and other methodologies (Table 1 & Figure 1). Those that are in comparatively high use in China—including, for example, the Modified Habitat Simulation method, the Monthly (yearly) Guarantee Rate method, the Tennant method, the Fish Habitat method and the Method of Minimum Flow in Dry

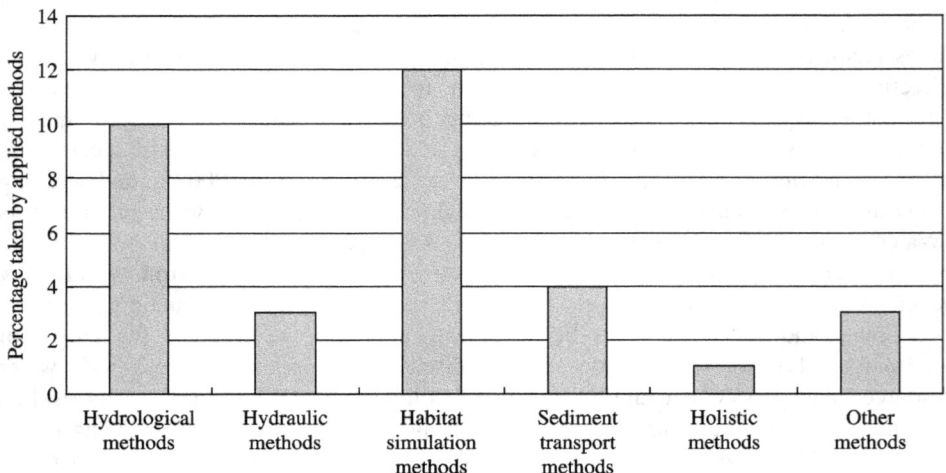

Figure 1. Application of different environmental flows methodologies in China.

Season—fall into two categories: habitat simulation methodologies and hydrological methodologies (Figure 1). Some geographical patterns in the use of environmental flow methods are apparent. For rivers of northern China where water resources are in short supply and pollution is often severe (e.g. the Yellow River, the Haihe or Hai River and the Liaohe or Liao River), hydrological methods have mainly been adopted. For rivers with relatively large flow volumes and better water quality (e.g. Songhua River and the Heilong River), methods linking habitat requirements and/or characteristics of target organisms are more commonly used. On the other hand, for rivers of mountainous regions such as the Yalong River and the Xiangxi River, hydraulic methods and habitat simulation methods have been more widely adopted.

Currently the parameters used in many methodologies are relatively simple. When developing relationships between river flows and organisms or their habitats, for example, most methods have utilized fish as the indicator species. As there is a lack of data on flow requirements for fish, some researchers have determined environmental flows for rivers in southern China with reference to the river velocity (e.g. 0.3–0.4 m/s) during fish spawning (Liu, 1999; Chen *et al.*, 2007). These methods require some validation that the flow is relevant for the particular species of interest.

Methodologies attempting to achieve a combination of outcomes are relatively rare. For instance, in the downstream reaches of the Yellow River, various methods have been applied to calculate water requirements for sediment transmission in the flood season, pollution control and maintaining river channel form. There is, however, still no model available to integrate these results into a single suite of flow recommendations. Methods such as the Contribution of Channel Divisions method (Zhang *et al.*, 2006) adopt an integrated approach, but simply rely on the highest demand amongst the different functions in recommending minimum flows. Consequently, there is often limited correlation between the recommended flows and the environmental objectives. Methodologies in which river flows and hydraulic variables are more directly linked to aquatic species characteristics and habitats have been the focus of increasing attention and are likely to contribute to a larger range of applications in the future.

Environmental Flows Application in Chinese Water Resources Management

Chinese water resources are distributed unevenly across the country, with about 80% of the nation's water resources in the Yangtze River Basin and its southern region. Around half the population, however, resides in these areas with 35% of the arable land and 55% of the GDP (Figure 2). In contrast, the region north of the Yangtze River Basin has 44% of the population, 59% of the arable land and 43% of the GDP but only 15% of the total water resources (Figure 2). In the Yellow River Basin, the Huai (Huaihe) River Basin and the Hai (Haihe) River Basin, water is in extremely short supply. This is reflected in the fact that the basins are home to around one-third of China's arable land, population and GDP, but only about 8% of the nation's water resources. In addition, variations in development at the local level result in different challenges for the allocation of environmental flows within each basin. These variations have created major challenges in providing environmental flows in some of the more water-scarce regions. In the Haihe and Liaohe Basins, critical water shortages mean that providing significant environmental flows in the short term is not a realistic option for water managers. Even in those parts where some

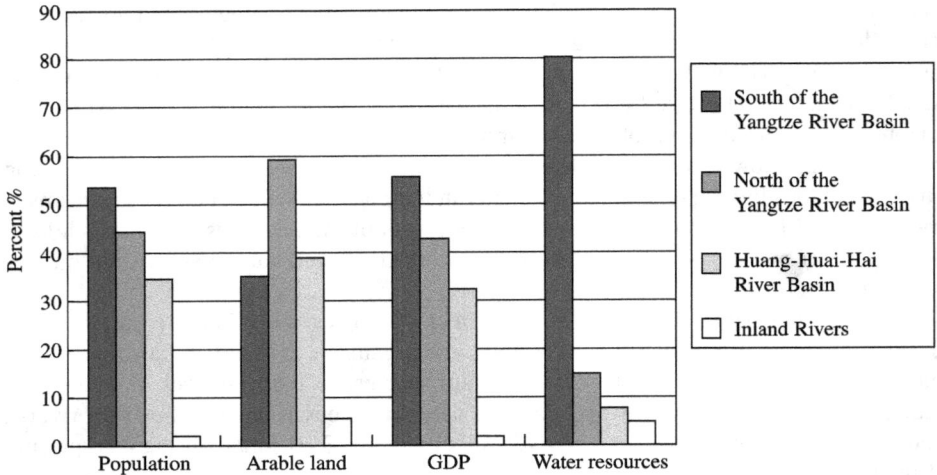

Figure 2. Distribution of water resources, population, arable land and GDP in China. *Source:* Liu & Chen (2001)

environmental flow has been provided (e.g. the trunk stream of the Huaihe River), water quality remains poor because of severe environmental pollution.

In other regions there has been more success in providing environmental flows. This has often been in response to particular (extreme) environmental challenges (e.g. the drying up of a river or lake, or severe pollution). The provision of water for the environment has included changes to operational rules (e.g. to require releases of water) as well as diversion projects, to increase flows in certain rivers or lakes.

In the Yellow and Tarim Rivers, environmental water releases have achieved favourable ecological results. In 2002, water releases from a number of large reservoirs were simultaneously carried out in the Yellow River. Repeated releases (eight times to date) have ensured that downstream reaches of the Yellow River have not dried up in the last five successive low-flow years. This has facilitated the recovery of downstream reaches of the Yellow River to some extent (Feng *et al.*, 2007). In the estuarine wetlands of the Yellow River Delta, 459 rare species were recorded in 2004, nearly double that of four years before, and the number of bird species increased from 187 in 1992 to 283 in 2004. During June and July of 2008, the eighth Yellow River environmental release took place, providing 3.6 billion m^3 of fresh water to inshore habitats, which should provide favourable conditions for fish migration and spawning in estuarine areas.

Since September 2000, through the implementation of a water release programme from Boisten Lake, 2.4 billion m^3 of water has been conveyed along the Tarim River to downstream reaches. At the same time, the average groundwater level in the valley has risen from -8.25m to -4.13m, (Ge *et al.*, 2008), and the spatial extent of groundwater has increased from a 450m wide belt adjacent to the river to more than one km wide. Consequently, the dominant species of tree (*Populus euphratica*) in downstream reaches of the Tarim River has expanded from a 200–250m-wide belt after the first water release to an 800m-wide belt (Ge *et al.*, 2008).

Elsewhere, improved environmental flows have seen lakes which were previously drying up now recovering. In 1992 the Dongjuyan Lake situated downstream along the

Heihe (Hei River) dried up entirely. Following subsequent environmental releases, however, a $23.66\,km^2$ lake appeared for first time in 10 years. A further eight environmental water releases has increased the lake's surface area to $35.7\,km^2$ (Qiao *et al.*, 2007). Around 0.25 billion m^3 of water has been transferred (on two occasions) from the Yellow River to Lake Baiyangdian, the largest freshwater lake in the Haihe Basin (Gong, 2008). The diversion increased the lake's surface area from $60\,km^2$ to $145\,km^2$ (Wang & Dai, 2007). The lake has subsequently reached its highest water level in 10 years and the number of fish and bird species in the lake has recovered significantly, as have as some endangered aquatic plants, such as *Euryale Ferox Salisb* and Water Chestnut (Wang & Dai, 2007). A similar water diversion project from the Yangtze River to Tai Lake has been used to transfer around 5 billion m^3 on four different occasions (2002, 2003, 2005 and 2006). This has allowed water managers to maintain water levels in the lake and surrounding rivers at 3.0–3.4m depth and has increased the flow in neighbouring rivers from 0–0.1m/s before the water diversion to 0.2–0.3m/s. The additional water has reduced water residence times in the lake by 50–100 days and reduced cyanobacteria blooms (Wu, 2008).

Discussion

Deficiencies in Flow Methodologies

Despite the shift in focus of environmental flows research to methods utilizing biological and hydraulic information, research is still only at the theoretical stage. This is largely because key information on aquatic organisms and their habitats is unavailable, relationships between river flows and China's aquatic biota are poorly understood, and analytical methods for establishing relationships between flow and ecological responses are in their infancy. As such, the use of more complex habitat simulation methods is restricted and these have generally only been used in rivers where there has been minimal human influence.

In addition, it is clear that although flow is an important factor influencing aquatic organisms and their habitats, it is not the sole factor in determining variations in biological populations and biomass. There exist many other influences, and particularly water quality remains an issue of great concern in China. Given this, a complete river monitoring system is required to provide the baseline data to support future environmental flows research. Improved data is needed on river hydrology, geomorphology, riparian condition, water quality and aquatic species.

The majority of rivers in northern China are highly silted, with 80% of the sediment load carried during the flood season. In these rivers sediment transportation has been a major focus for water managers, and the water required to balance incoming and outgoing sediment loads during the flood season has been a focal point of environmental flows research. This requires consideration of not only the flow required to remove sand, but also impacts of different flows on sediment washing and how sediments will accumulate based on runoff and other conditions in the upstream catchment. As such, calculation methods have become increasingly complex and while various methods have been proposed from recent research, most calculations are based on hydrological statistics and an accepted method has not yet been agreed.

River pollution is arguably the biggest challenge to both research on and provision of environmental flows in China. There remain many divergent views on whether the water to dilute pollution should be regarded as 'environmental flows' and the calculation methods available generate extremely variable results. Consequently, further research is required into how to consider water quality issues with other protection targets.

One of the biggest shortfalls of China's environmental flows research is the lack of integrated calculation methods. The recommended environmental flow is often targeted to a single outcome, and most methods provide flow recommendations at a relatively crude time step (such as monthly, annually, non-flood season and flood season). This doesn't represent the natural variability of river flows adequately and consequently is unlikely to yield the flows necessary to provide for the identified ecological assets.

Whilst substantial research on the floodplain wetlands of China has been carried out in recent years, most researchers do not link riverside wetlands with environmental flows, but merely estimate wetland water requirements through water balance equations using evaporation and precipitation estimates (Wang, L. Q. *et al.*, 2008). The importance of lateral connectivity between wetland and floodplain habitats and the river systems has thus been neglected despite international recognition of the importance of connectivity (e.g. Bunn & Arthington, 2002). Although some scholars have begun to study suitable flooding flows, they focus primarily on the high flow or flood conditions needed to maintain channel form and prevent channel shrinkage (Liu, X. Y. *et al.*, 2007). It should be noted that to maintain river health and the health of associated floodplains and wetlands, it is necessary to provide a range of flows that mimic essential components of the natural flow regime including low flows and floods (Arthington *et al.*, 2006). Research should therefore continue on appropriate calculation methods, particularly those calculations that attempt to integrate multiple objectives to achieve multiple outcomes.

Obstacles to the Implementation of Environmental Flows

(1) Water requirements outside the river course. Human water use, including the exploitation of groundwater, is directly impacting the delivery of environmental flows, particularly during periods of low and extremely low flow. At present, surface water resource utilization in most river basins of China exceeds 60% (e.g. the Haihe, the Yellow River, the Huaihe and the Liaohe) (Wang & Zhang, 2008a). This situation is likely to be exacerbated in the future. For example, for the Yellow River, estimates are that the water required to maintain river health and natural functions is $23-26$ billion m^3, while water requirements for social and economic development will reach 40 billion m^3 by 2030 (Liu & Liu, 2008). The natural runoff volume of the Yellow River, however, is only about 52 billion m^3. It is quite apparent both demands cannot be satisfied (Liu & Liu, 2008). In addition, in most areas of China precipitation falls mainly between June and September, whilst crop irrigation is required in April and May. Seasonal differences between precipitation and crop water requirements (and the resultant releases to provide irrigation water) have altered natural flow patterns, and present a serious obstacle to the implementation of environmental flows. Improving environmental flows will require reductions in abstractions, which in turn will require improvements in water allocation planning and the water licence system.

(2) River pollution. Despite the provision of environmental flows to some rivers, the desired ecological objectives often cannot be achieved due to the severe pollution levels in many Chinese rivers. The combination of a shortage of water together with water quality

degradation exists in all seven of China's major river basins (Wang & Zhang, 2008b). In the Huaihe, for example, despite the provision for water for environmental flow purposes, water quality remains poor. As such, those implementing environmental flows need to consider, and if possible address, both water quantity and quality issues. Clearly, it is necessary to reduce exploitation and utilization rates of water resources and, critically, take additional measures to control the discharge of pollutants into rivers.

(3) Dams, reservoirs and other WRD infrastructures. Environmental benefits can be achieved by changing the operation of dams and other water storage infrastructure without necessarily altering the water volume allocated to environmental flows. Most Chinese rivers have been subject to the construction of dams, reservoirs and flood protection infrastructure (dykes, levees etc). This has altered physical habitats, disrupted longitudinal connections downstream and lateral connectivity with wetlands and floodplain habitats, and changed natural flow patterns. For example, the Sanmenxia Reservoir on the Yellow River has altered the seasonal pattern of downstream runoff, removing much of the natural seasonal variation and resulting in a relatively even flow downstream throughout the year (Guo & Yang, 2005). At present, most WRD projects are designed to provide only minimal environmental releases, with 10% of the natural runoff regarded as the standard minimum environmental flow release, based on the Tennant method (Tennant, 1976).

Researchers have shown that there is a close relationship between the operational pattern of the Sanmenxia Reservoir and incidences of no flow in the main channel (Wei *et al.*, 2004, Guo & Yang, 2005). Thus the number of dry days in the downstream reaches of the Yellow River can be reduced by controlling the operational pattern of the reservoir, particularly by increasing releases from March to June during spring irrigation (Wei *et al.*, 2004).

Modifying dam operational patterns to provide environmental flows can be constrained by other management objectives (e.g. hydropower generation) and existing physical constraints (for example, nature of existing dam outlets). Particular consideration should be given to incorporating structural components (such as variable offtakes from dams) into the design of new dams because there is often limited capacity to alter existing structures and retrofitting is expensive (Richter & Thomas, 2007).

International experience shows that there is scope to significantly improve environmental flows through modification of operational rules for hydropower stations. Between 1991 and 1996, the American Tennessee Valley Authority (TVA) used minimum flow requirements (based on aquatic habitat, water supply and waste assimilation) and specific dissolved oxygen targets to optimize the control and operational patterns for 20 reservoirs (Higgins & Brock, 1999). In 1996 the American Federal Energy Regulatory Commission required reservoir operational plans be revised in light of potential ecological and environment influences, including raising the minimum release flows, increasing or improving fish passage, providing for periodic large releases and ensuring conservation of riparian habitats (National Park Service, 1996).

Researchers and water managers in China have paid attention to events such as those described above. However, because of China's broad range of climates—including tropic, subtropic, temperate and alpine zones—there are great variations in hydrology, hydraulics and biological habitats within and between China's rivers. Developing environmental flow methodologies that are applicable across a country with such a diverse geography is likely to be a long and challenging process.

Conclusions

Research into environmental flows in China can be divided into four stages: a stage of preliminary recognition in the 1970s, a stage of exploration in the 1980s, a stage of rapid development in theory and research from the 1990s to the beginning of the 21st century and, most recently, a stage from the beginning of the 21st century during which methods have been increasingly applied and tested.

Improvements have been made to the methods for calculating environmental flows and research has provided environmental flows with a strong theoretical foundation. However, because relationships between flow and ecology are complex, research on environmental flows is still largely in the exploratory stage and assessment methodologies require refinement. Research into biological and hydraulic methods is a current focal point and appears to be the direction of future studies. To improve these methods it will be necessary to collect and collate hydraulic, water quality and biological information, in order to provide basic data for establishing relationships between flow and river ecology.

Although environmental flows have been implemented in several river basins and with some improvement of ecological condition, the provision of environmental flows remains a challenge. With social and economic development and a growing population, water resources are likely to come under greater pressures and demands and conflicts between uses and users are likely to sharpen. Improving the health of China's rivers presents an array of challenges to water managers and researchers which will require further research and debate. There is a need to increase the efficiency of water use and to reduce water abstractions, and the discharge of pollutants needs to be reduced and flows need to be provided to meet ecological needs. Ultimately, there is a need to harmonize ecological water requirements with economic water demands. Water needs to be made available to protect river ecosystems while supporting economic development. Balancing the water requirements of humans and river ecosystems will no doubt be a focal point for future research and arguably presents China's biggest environmental challenge.

Acknowledgements

The authors would like to thank Mr Robert Speed for valuable suggestions throughout the writing of this paper. We are also grateful to Dr Shen Dajun, Dr Liu Bin and other reviewers for their thoughtful comments and revisions to this paper. This paper is the result of a project undertaken under the auspices of the Australian Department of the Environment, Water, Heritage and the Arts and the Chinese Ministry of Water Resources, with funding provided by AusAID, the Australian Agency for International Development.

References

Arthington, A. H., Bunn, S. E., Poff, N. L. & Naiman, R. J. (2006) The challenge of providing environmental flow rules to sustain river ecosystems, *Ecological Applications*, 16, pp. 1311–1318.
Bunn, S. E. & Arthington, A. H. (2002) Basic principles and ecological consequences of altered flow regimes for aquatic biodiversity, *Environmental Management*, 30, pp. 492–507.
Chang, B. Y., Xue, S. G. & Zhang, H. Y. (1998) *Rational Utilization and Optimizing Regulation of Water Resources in the Yellow River* [in Chinese] (Zhengzhou: Press of Yellow River).
Chen, M. J., Feng, H. L., Wang, L. Q. & Chen, Q. Y. (2007) Calculation methods for appropriate ecological flow, *Advances in Water Science* [in Chinese], 18(5), pp. 745–751.
Chen, Y. N., Hao, X. M. & Li, W. H. (2008) An analysis of the ecological security and ecological water requirements in the inland river of arid region, *Advance in Earth Science* [in Chinese], 23(7), pp. 732–738.

Dispatch Act of Water Quantity from the Yellow River (2006) Law applied since 1 August 2006 (Beijing: China State Department).

Fang, Z. Y. (1988) *Protection Notebook of Water Resources* [in Chinese] (Nanjing: Hehai University Press).

Feng, J. C., Ran, D. P., Wang, X. G., Wang, Y. H. & Liu, J. (2007) Preliminary appraisal on implementation effect of integrated water regulation of the Yellow River, *Yellow River* [in Chinese], 29(6), pp. 1–3.

Fu, X. C., Tang, T., Jiang, W. X., Li, F. Q., Wu, N. C., Zhou, S. C. & Cai, Q. H. (2008a) Impacts of a small hydropower plant on macro invertebrate communities, *Acta Ecologica Sinica* [in Chinese], 28(1), pp. 45–53.

Fu, X. C., Wu, N. C., Zhou, S. C., Jiang, W. X., Li, F. Q. & Cai, Q. H. (2008b) Impacts of a small hydropower plant on macroinvertebrate habitat and an initial estimate for ecological water requirement of Xiangxi River, *Acta Ecologica Sinica* [in Chinese], 28(5), pp. 1942–1943.

Ge, X. Q., Wu, J. & Yu, L. (2008) Impact post assessment of emergent water conveyance project of ecological based at downstream the Tarim River, *HongShui River* [in Chinese], 27(1), pp. 58–62.

Gippel, C. J., Bond, N. R., James, C. & Wang, X. (2009) An asset-based, holistic, environmental flows assessment approach, *International Journal of Water Resources Development*, 25(2), pp. 301–330.

Gong, Z. H. (2008) The Lake Baiyangdian has reached the highest water level in recent ten years, Xinhua News Agency, 17 July. Available at http://www.he.xinhuanet.com/zhuanti/2008-07/18/content_13855866.htm

Guo, Q. Y. & Yang, Z. F. (2005) Post-project ecological analysis for the Sanmenxia Dam, *Acta Scientiae Circumstantiae* [in Chinese], 25(5), pp. 580–585.

Guo, W. X. & Xia, Z. Q. (2007) Study on ecological flow in the middle and lower reaches of the Yangtze River, *Journal of Hydraulic Engineering* [in Chinese], 10, pp. 618–623.

Hao, Z. C. & Shang, S. H. (2008) Multi-objective assessment method based on physical habitat simulation for calculating ecological river flow demand, *Journal of Hydraulic Engineering* [in Chinese], 39(5), pp. 557–563.

Higgins, J. M. & Brock, W. G. (1999) Overview of reservoir release improvement at 20 TVA dams, *Journal of Energy Engineering*, 125(1), pp. 1–17.

Jia, J. S., Yuan, Y. L. & Li, T. J. (2003) *Statistics of Dams in China in 2003* (Beijing: China's Commission on Dams) [in Chinese]. Available at: http://www.chincold.org.cn/zt/dams/2003zg.asp (accessed 20 September 2008).

Li, L. J. & Zheng, H. X. (2000) Environmental and ecological water consumption of river systems in Haihe–Luanhe Basins, *Acta Geographica Sinica* [in Chinese], 55(4), pp. 495–500.

Liu, C. M. & Chen, Z. K. (2001) *Assessment of Chinese Water Resources Status and Analysis on Developing Trends of Supply and Demand* [in Chinese] (Beijing: China Water Conservancy and Hydroelectricity Press).

Liu, C. M. & Liu, X. Y. (2008) Healthy river: Essence and indicators, *Acta Geographica Sinica* [in Chinese], 63(7), pp. 683–692.

Liu, C. M., Men, B. H. & Song, J. X. (2007) Ecological hydraulic radius approach for estimating instream ecological flow requirements, *Progress in Natural Science*, 17(1), pp. 42–49.

Liu, C. M., Wang, L. X. & Xia, J. (2004) *Research on Regional Allocation of Environment Construction and Ecological Water Requirement in Northwest China* [in Chinese] (Beijing: Science Press).

Liu, C. M., Xia, J. & Yu, J. J. (2006) *The Water and Ecology-Environment Problems in North East Region and the Research of Protect Countermeasure* [in Chinese] (Beijing: Science Press).

Liu, J. K. (1999) *Advanced Hydrobiology* [in Chinese] (Beijing: Science Press).

Liu, J. K. & Yu, Z. T. (1992) Water quality changes and effects on fish populations in the Hanjiang River, China, following hydroelectric dam construction, *Regulated Rivers: Research & Management*, 7, pp. 359–368.

Liu, S. X., Mo, X. G., Xia, J., Liu, C. M., Lin, Z. H., Men, B. H. & Ji, L. N. (2006) Uncertainty analysis in estimating the minimum ecological instream flow requirements via wetted perimeter method: Curvature technique or slope technique, *Acta Geographica Sinica* [in Chinese], 61(3), pp. 273–281.

Liu, S. X., Xia, J., Mo, X. G., Lin, Z. H., Liu, C. M., Xuan, X. B. & Wu, C. H. (2007) Estimating ecological instream flow requirements for the donating rivers in the western route South to North Water Transfer project in China based on the relationship between the life habit and flow variation, *South-to-North Water Transfers and Water Science & Technology* [in Chinese], 5(5), pp. 12–19.

Liu, X. Y., Li, T. H., Zhao, Y. A., Jin, L. & Ni, J. R. (2002) Water demand for sediment transport in the lower Yellow River, *Journal of Basic Science and Engineering* [in Chinese], 10(3), pp. 253–262.

Liu, X. Y., Shen, G. Q., Li, X. P. & Zhang, Y. F. (2007) Flood volume required for keeping the main channel from shrinkage in the lower Yellow River, *Journal of Hydraulic Engineering* [in Chinese], 38(9), pp. 1140–1145.

Liu, Y. H. (2000) *Rational Utilization of Water Resources and Protection of Ecosystem Environment in Chaidamu Basin* [in Chinese] (Beijing: Science Press).

Men, B. H., Liu, C. M., Xia, J., Liu, S. X. & Ji, L. N. (2005) Estimating and evaluating on minimum ecological flow of western route project of China's South-to-North Water Transfer Scheme for water exporting rivers, *Journal of Soil and Water Conservation* [in Chinese], 19(5), pp. 135–138.

Ministry of Environmental Protection of the People's Republic of China (2007) *China's State of the Environment Report, 2006* [in Chinese]. Available at: http://www.zhb.gov.cn/ztbd/sjhjr/2007hjr/tpbd56/200706/P020070625532626111313.pdf. (accessed 20 September, 2008).

National Park Service (1996) *River Renewal: Restoring Rivers through Hydropower Dam Relicensing* (Washington, DC: National Park Service).

Nehring, R. (1979) *Evaluation of Instream Flow Methods and Determination of Water Quantity Needs for Streams in the State of Colorado* (Fort Collins, CO: Colorado Division of Wildlife).

Nelson, F. A. (1984) *Guidelines for Using the Wetted Perimeter (WETP) Computer Program of the Montana Department of Fish* (Bozeman, MT: Montana Department of Fish, Wildlife and Parks).

Qian, Z. Y. & Zhang, G. D. (Eds) (2001) *Strategic Study on Sustainable Water Resource Development in China* [in Chinese] (Beijing: China Water and Power Press).

Qiao, X. X., Jiang, X. H., Chen, J. N., Yin, H. J. & Chen, L. (2007) Effect of transferring water on ecological environment in east and west Juyanhai lake at the lower reaches of Heihe River, *Journal of Northwest Agriculture & Forestry University (Natural Science Edition)* [in Chinese], 35(6), pp. 190–195.

Recycled Water of Huai River Engineering Ltd. and Planning and Design Institute (2006) *Environmental Impact Assessment of River Basin Planning (SL45-92)* (Beijing: China WaterPower Press).

Richter, B. D. & Thomas, G. A. (2007) Restoring environmental flows by modifying dam operations, *Ecology and Society*, 12(1), pp. 1–12.

Shen, G. Q. (1988) Water conservancy encyclopaedia China, branch entry of "environmental hydro-engineering" (discussion paper), *Protection of Water Resources* [in Chinese], 1, pp. 5–8.

SKM, CRC for Freshwater Ecology, Freshwater Ecology (NRE) & Lloyd Environmental Consultants (2002) *The FLOWS Method—A Method for Determining Environmental Water Requirements in Victoria*. Report prepared by Sinclair Knight Merz, the Cooperative Research Centre for Freshwater Ecology, Freshwater Ecology (NRE) and Lloyd Environmental Consultants (Melbourne: Department of Natural Resources and Environment). Available at http://www.envict.org.au/file/Flows_Methodology.pdf (accessed 24 September 2008).

Song, J. X. & Li, H. E. (2004) *Research on Eco-Environment Water Requirements in Weihe River* [in Chinese] (Beijing: Publishing House of Irrigation and Power).

Sun, T. & Yang, Z. F. (2005) Study on the methods for quantifying the environmental flow in estuaries, *Acta Scientiae Circumstantiae* [in Chinese], 25(5), pp. 573–580.

Sun, Y., Shao, D. G. & Gu, W. Q. (2008) Calculation approach of ecological water demand based on breeding of critical species in the middle reaches of Hanjiang River, *South-to-North Water Transfers and Water Science & Technology* [in Chinese], 6(3), pp. 97–101.

Tennant, D. L. (1976) Instream flow regimens for fish, wildlife, recreation and related environmental resources, *Fisheries*, 1(4), pp. 6–10.

Tharme, R. E. (2003) A global perspective on environmental flow assessment: Emerging trends in the development and application of environmental flow methodologies for rivers, *River Research and Applications*, 19, pp. 397–441.

Tharme, R. E. & King, J. M. (1998) *Development of the Building Block Methodology for Instream Flow Assessments and Supporting Research on the Effects of Different Magnitude Flows on Riverine Ecosystems*, Freshwater Research Unit WRC Report no. 576/1/98 (Cape Town, South Africa: University of Cape Town).

Wan, D. H. & Xia, J. (2007) Analysis of ecological water requirements in Yalong River of west route of South-to-North Water Transfer Project, *Engineering Journal of Wuhan University* [in Chinese], 40(6), pp. 1–4.

Wang, J., Bao, W. K., Pang, X. Y., Fan, J. R. & Yang, A. Q. (2006) Study on ecological water requirement of dry valleys in the upper reaches of the Dadu River, *Journal of Natural Resources* [in Chinese], 21(2), pp. 252–259.

Wang, L. M. & Dai, Y. (2007) Environment impact assessment of river water diversion from Yellow River to Baiyangdian Lake, *Haihe River Hydraulic Engineering* [in Chinese], 7, pp. 12–15.

Wang, L. Q., Chen, M. J., Dai, X. Q., Feng, H. L., Wang, G. X. & Huang, C. S. (2008) Analysis on ecological water dam and eco-hydrological configuration of wetlands in Songliao Basin, *Acta Ecologica Sinica* [in Chinese], 28(6), pp. 2894–3000.

Wang, X. Q. (2002) Research on Environmental Water Requirement in Huang-Huai-Hai Plain [in Chinese] unpublished report of post-doctoral stations in Beijing Normal University.

Wang, X. Q. (2007) *Theories, Methods and Applications of River Environment Flow Requirement* [in Chinese] (Beijing: Publishing House of Irrigation and Power).

Wang, X. Q. & Zhang, Y. (2006) Discussion on ecological water demand in integrated viewpoint of water quantity and quality, *Environmental Protection* [in Chinese], 2B, pp. 75–78.

Wang, X. Q. & Zhang, Y. (2008a) The allowable exploitation rate of river water resources of the seven major rivers in China, *Journal of Natural Resources* [in Chinese], 23(3), pp. 500–506.

Wang, X. Q. & Zhang, Y. (2008b) Situation evaluation on river channel ecological water requirements of seven major basins in China, *Journal of Natural Resources* [in Chinese], 23(1), pp. 95–102.

Wang, X. Q., Liu, C. M. & Yang, Z. F. (2001a) Method of resolving lowest environmental water demands in river course (I), *Acta Scientiae Circumstantiae* [in Chinese], 21(5), pp. 544–547.

Wang, X. Q., Liu, C. M. & Yang, Z. F. (2001b) Method of resolving lowest environmental water demands in river course (II), *Acta Scientiae Circumstantiae* [in Chinese], 21(5), pp. 548–552.

Wang, X. Q., Zhang, Y. & Liu, C. M. (2003) Study of the basic environmental water requirement of the rivers in Huang-Huai-Hai Plain, *Geographical Research* [in Chinese], 22(2), pp. 169–176.

Wang, X. Q., Zhang, Y. & Liu, C. M. (2007a) Water quantity-quality combined evaluation method for rivers' water requirements of the instream environment in dualistic water cycle: A case study of Liaohe River Basin, *Journal of Geographical Sciences*, 16(3), pp. 304–316.

Wang, X. Q., Zhang, Y. & Liu, C. M. (2007b) Estimation of eco-water requirement in the Liaohe River Basin, *Geographical Research* [in Chinese], 2(1), pp. 22–28.

Water Resources Protection Bureau of Haihe River Basin (2000) Report of Environment Water Utilization in Haihe River Basin, unpublished report [In Chinese].

Wei, H. L., Ni, J. R. & Wang, Y. D. (2004) Impact of operation mode of Sanmenxia Project on ecological environment in lower reaches of Yellow River, *Journal of Hydraulic Engineering* [in Chinese], 9, pp. 9–17.

WET (2007) *Water Entitlements and Trading Project (WET Phase 2) Final Report* December 2007 [in English and Chinese] (Beijing: Ministry of Water Resources, People's Republic of China and Canberra: Department of the Environment, Water, Heritage and the Arts, Australian Government). Available at: http://www.environment.gov.au/water/action/international/wet2.html

World Commission on Dams (WCD) (2000) *Dams and Developments: A New Framework for Decision-Making* Report of the World Commission on Dams (Cape Town: WCD).

Wu, C. H. (2007) The habitat simulation method of research on the ecological flow in river course of Yalong main stream, *Ecological Science* [in Chinese], 26(6), pp. 536–539.

Wu, H. Y. (2008) Practice and exploration on maintenance of river-lake ecosystem health for Taihu Basin with project of water diversion from Yangtze River to Taihu Lake, *Irrigation and Power Technology* [in Chinese], 39(7), pp. 4–8.

Xu, Z. X., Chen, M. J. & Dong, Z. C. (2004) Research on the calculation method of river minimum eco-environment water based on the analyses of ecosystem, *Irrigation and Power Technology* [in Chinese], 35(12), pp. 14–18.

Xu, Z. X., Wang, H. & Chen, M. J. (2005) Research on the calculation method of river minimum eco-environment water based on the analyses of ecosystem, *Irrigation and Power Technology* [in Chinese], 36(1), pp. 31–34.

Yan, D. H., He, Y., Deng, W. & Wang, J. D. (2001) Ecological water demand by river system in East Liaohe River Basin, *Journal of Soil and Water Conservation* [in Chinese], 15(1), pp. 46–49.

Yang, S. M., Shao, D. G. & Shen, X. P. (2005) Quantitative approach for calculating ecological water requirement of seasonal water deficient rivers, *Journal of Hydraulic Engineering* [in Chinese], 36(11), pp. 1341–1346.

Yang, Z. F., Cui, B. S., Liu, J. L., Wang, X. Q. & Liu, C. M. (2003) *The Theory, Method and Application of Eco-Environment Water* [in Chinese] (Beijing: Science Press).

Yangtze River Water Resources Protection Science Research Institute (1979) *Preliminary Exploration of Environmental Water Requirement* (R) [in Chinese].

Zhang, W. G., Huang, Q. & Jiang, X. H. (2008) Study on instream ecological flow based on physical habitat simulation, *Advances in water science* [in Chinese], 19(2), pp. 192–198.

Zhang, Y. (2003) Research on the ecological water requirements of rivers and plant in the Yellow River [in Chinese]. Doctoral thesis, Beijing Normal University.

Zhang, Y., Yang, Z. F. & Wang, X. Q. (2006) Methodology to determine regional water demand for instream flow and its application in the Yellow River Basin, *Journal of Environmental Sciences*, 18(5), pp. 1031–1039.

Zhang, Y., Zheng, B. H., Wang, X. Q. & Hu, C. H. (2007) Study of ecological instream basin of the Hun River and Taizi River, *Acta Scientiae Circumstantiae* [in Chinese], 27(6), pp. 937–944.

Zhang, Y. M., Hu, S. J., Zhai, L. X. & Shen, B. (2008) Water requirement for the self purification of the Tarim River, *Journal of Lanzhou University (Natural Science)* [in Chinese], 44(2), pp. 22–28.

Zhao, C. S., Liu, C. M., Xia, J., Wang, G. S., Liu, Y., Sun, C. L., Wang, R. & Ji, X. Y. (2008) Instream ecological flow of dammed river: A case study of Huaihe River, *Journal of Natural Resources* [in Chinese], 23(3), pp. 400–412.

Zheng, J. P., Wang, F., Hua, Z. L. & Zhu, J. D. (2005) Research on ecological water requirement of Haihe River Estuary, *Journal of Hehai University (Natural Science)* [in Chinese], 33(5), pp. 518–521.

Zhong, Y. G. & Power, G. (1996) Environmental impacts of hydroelectric projects on fish resources in China, *Regulated Rivers: Research & Management*, 12, pp. 81–98.

Appendix

Table 1. Environmental flow methodologies applied in China

Method category	Methodology	Brief method description	Example application(s)	References
Hydrological methods	Tennant or Montana Method (Tennant, 1976)	A hydrological method using historical flow data to calculate a percentage of average annual flow – 10% is often used as the minimum environmental flow requirement.	Yellow River, Haihe, Huaihe, Yalong River, Liaohe	Wang et al. (2003); Yang et al. (2003); Liu C. M. et al. (2006); Men et al. (2005)
	Monthly (yearly) Guarantee Rate (Wang et al., 2003)	A hydrological method adaptation from the Tennant method (Tennant, 1976) using historical flow data to calculate a minimum monthly flow (as a percentage). Various grades or guarantee rates are attributed to different flow percentages.	Yellow River, Haihe, Huaihe River, Liao River, Songhua River	Wang et al. (2003); Yang et al. (2003); Liu C. M. et al. (2006)
	Average Flow in the Lowest Flow Season method (Water Resources Protection Bureau of Haihe River Basin, 2000)	A hydrological method adapted from the 7Q10 method (the lowest consecutive seven day stream flow that occurs in a ten year period). In China this is modified to the driest monthly flow in the last ten years or the 90% duration driest monthly flow.	Haihe River	Water Resources Protection Bureau of Haihe River Basin (2000)
	Minimum Monthly Field Runoff method (Li & Zheng, 2000)	A hydrological method in which flow recommendations are calculated from the yearly average of the minimum monthly field runoff.	Haihe, Liaohe, Dadu River	Li & Zheng (2000); Yan et al. (2001); Wang J. et al. (2006)
	Minimum Flow in Low Flow Season (Wang et al., 2007b)	A hydrological method using the minimum runoff volume in low flow season as the flow recommendation for the dry season.	Liaohe, Hunhe, Taizihe River, Songhua River	Wang et al. (2007b); Liu C. M. et al. (2006)

Table 1. *Continued*

Method category	Methodology	Brief method description	Example application(s)	References
	Contribution of Channel Division method (Zhang *et al.*, 2006)	A method intended to achieve a combination of outcomes (ecological, pollution dilution and sediment transmission). Ecological requirements are based upon the hydrological method of Wang (Wang *et al.*, 2003). Rivers are divided into functional sections and may be attributed more than one function (for example: ecological, pollution dilution and sediment transport functions). Requirements in different reaches are integrated within a whole river basin such that flows from upper reaches are taken into account in the calculation of requirements in lower reaches.	Yellow River	Zhang *et al.* (2006)
	Function method (Wang *et al.*, 2001a)	A method in which rivers are divided into sections. The section's lowest environmental water demands are based on each sections upstream water quality aims.	Weihe	Wang *et al.* (2001b); Song & Li (2004)
	Pollutants-Flow Curve method (Chang *et al.*, 1998)	A method based upon empirical relationships between pollutants and river flows. Flow recommendations are based upon the discharge required to attain various water quality aims or targets.	Yellow River	Chang *et al.* (1998)
Hydraulic methods	Wetted Perimeter method (Xu *et al.*, 2004; Liu S. X. *et al.*, 2006)	A hydraulic method based on the Wetted Perimeter method (Nelson, 1984). Points of inflection in the relationship between the proportion of the perimeter of the stream that is wet and discharge are used to define an optimal discharge for biota of interest.	Liaohe, Songhua River, Ying river tributary of Huai River, Yalong River	Chen *et al.* (2007); Xu (2005); Liu S. X. *et al.* (2006)

Table 1. *Continued*

Method category	Methodology	Brief method description	Example application(s)	References
	Ecological Hydraulic Radius Method (Liu S. X. *et al.*, 2007b)	A hydraulic method modified from the R2Cross method (Nehring 1979) in which flow recommendations are derived from flows necessary to meet specific hydraulic criteria (such as mean depth, % wetted perimeter and mean stream velocity) usually performed for shallow water or riffle habitats.	Yalong River	Liu S. X. *et al.* (2007); Wan & Xia (2007)
	Water Quality Model method (Zhang, 2003)	A method based on the analytic equations of hydraulic, pollutant dilution and pollutant diffusion. The river dilution and self-purifying targets are used to define flow recommendations.	Yellow River, Tarim River, Han River	Zhang (2003); Zhang Y. M. *et al.* (2008); Yang *et al.* (2005)
Habitat simulation methods	Modified Habitat Simulation method (Zhang *et al.*, 2007; Zhang W. G. *et al.*, 2008; Hao & Shang, 2008)	A habitat simulation method in which relationships between flow velocity and aquatic species habitats are used to develop flow recommendations.	Yalong River, Yellow River, Yangtze River, Hun River, Taizi River, Han River, Xiangxi River	Wu (2007); Zhang W. G. *et al.* (2008); Hao & Shang (2008); Fu *et al.* (2008b); Zhang *et al.* (2007); Guo & Xia (2007); Sun *et al.* (2008); Zhao *et al.* (2008)
	Fish Habitat Method (Xu, 2005; Chen *et al.*, 2007)	A habitat simulation method in which relationships between various hydraulic parameters (water depth, flow velocity) and fish spawning requirements are used to develop flow recommendations.	Yinghe tributary, Shawo River of the Huaihe, Liaohe, Songhua River	Xu (2005); Chen *et al.* (2007)
	Aquatic Productivity method (Chen *et al.*, 2007)	A method based on empirical relationships between aquatic productivity (usually fish) and flow.	Songhua River, Heilong River	Chen *et al.* (2007)

Table 1. *Continued*

Method category	Methodology	Brief method description	Example application(s)	References
	Life-Habit-Flow-Variation method (Liu C. M. *et al.*, 2007)	A method based on the establishment of quantitative relationships between flow variation and habit and/or characteristics of target organisms. Critical months of key life history stages are identified for target organisms and recommendations are based on median flows and coefficient of variation for the corresponding month.	Yalong River	Liu S. X. *et al.* (2007)
Sediment transport methods	Minimum Sediment Transmission Volume in Flood Seasons method (Li & Zheng, 2000)	A calculation based upon the ratio of the years' average sediment transmission volume over the average value of the years' maximum monthly average change in sediment load.	Haihe, Liaohe	Li & Zheng (2000); Liu C. M. *et al.* (2006)
	Balanced Sediment Transmission method (Liu *et al.*, 2002)	A calculation method based upon the flow required to balance incoming and outgoing sediment loads.	Yellow River	Liu *et al.* (2002)
	Experience method (Wang, 2002; Zhang, 2003)	A calculation method based on the historical capacity of different flows in flood seasons and non-flood seasons to transmit sediments.	Yellow River, Haihe, Huaihe	Wang (2002); Zhang (2003)
Holistic methods	FLOWS method (SKM *et al.*, 2002)	This method is a derivative of the bottom-up, prescriptive BBM (Tharme & King, 1998). It uses a framework with a series of logical steps undertaken by a small panel of scientific experts	Jiao River (Zhejiang)	Gippel *et al.* (2009)
Other methods	Salinity Simulation method (Zheng *et al.*, 2005; Sun & Yang, 2005)	A method based on of the relationship between salinity distributions under different flows association with fish habitats. Flow recommendations are dependent upon fishes' requirements and salinity distributions.	Haihe, Yellow River	Zheng *et al.* (2005); Sun & Yang (2005)
	Groundwater method (Liu *et al.*, 2004)	A method based on relationships between groundwater levels and eco-physiological responses of vegetation.	Tarim River	Liu *et al.* (2004)

An Asset-based, Holistic, Environmental Flows Assessment Approach

CHRISTOPHER J. GIPPEL, NICK R. BOND, CASSANDRA JAMES &
XIQIN WANG

ABSTRACT   *This paper describes a site-based, ecological asset-based, holistic environmental flows assessment approach, and demonstrates its application to reaches of the Jiaojiang (Jiao River) Basin, Taizhou, Zhejiang Province, the People's Republic of China. The methodology broadly combines information on ecological and other assets associated with the river system (in this case, fish, vegetation, water quality and geomorphology) together with information that links these assets to aspects of the flow regime via hydraulic relationships. This is a site-based methodology, and it requires a medium-level effort and budget. The methodology hinges on being able to gain a basic understanding of the ecology and geomorphology of the stream system, having daily flow series' available, and having the capacity to develop hydraulic models. A comparison of the flow regimes recommended for the Jiaojiang reaches with recommendations derived from two hydrology-only methods found little correspondence. This was explained by the failure of hydrology-only methodologies to take into account the downstream change in the relationship between a river's geomorphic and hydrologic characteristics (i.e. expressed as hydraulics). Also, the ecological assumptions made by the hydrology-only methods cannot necessarily be applied in a generic way.*

Introduction

From the policy perspective, management of river flows for sustainability is usually influenced by a combination of economic, social and environmental (triple bottom line) considerations. Science-based recommendations concerning environmental flows are usually made, at least initially, in isolation from the social and economic issues. However, the scientific work is not undertaken within a policy vacuum, as the results of environmental flows assessments need to be compatible with the overall strategy being used to manage rivers, floodplains and associated wetlands. The 11th Five-Year Plan of the People's Republic of China (for the period 2006–10) requires the establishment of a water rights allocation and transfer system, in addition to improving the water resources compensation system. This will likely involve increased consideration of the environmental flows needs of rivers in water resource planning. Although the concept of environmental flows is not new in the People's Republic of China (Wang, 2006; Jiang *et al.*, 2006; Jiang, 2007), in general, there remains little use of theoretical models and quantitative methods (Jiang *et al.*, 2006). There is a need then to develop and test environmental flows methodologies appropriate to the water resources situation that applies in the People's Republic of China, which can be briefly described as geographically highly variable, with many rivers coming under very high and rapidly increasing levels of demand. This paper describes a site-based, ecological asset-based, holistic environmental flows assessment approach, and demonstrates its application to reaches of the Jiaojiang (Jiao River) Basin in Taizhou, Zhejiang Province, People's Republic of China (Figure 1).

River environmental management policies generally fall into the categories of tenure-based, issues-based, ecological asset-based or process-based. Tenure-based policies consider the ownership of the water, or the land through which the river flows, to be of fundamental importance; issues-based policies focus on managing particular identified problems, such as salinity, riparian vegetation, pest-species or fish passage;

Figure 1. Location of Taizhou study area in eastern Zhejiang Province, People's Republic of China.

ecological asset-based policies focus on protecting key identifiable assets such as biodiversity, threatened species, native species, species of high conservation value, certain habitats or the relative health of ecosystems; and process-based policies focus on maintaining or restoring the physical, chemical and biological processes that sustain ecological assets. Examples are nutrient cycling, water flows, hydraulics, sediment dynamics, dispersal, adaptation, disturbance and functional interactions. The environmental flows method described in this paper suits an asset-based policy approach, and other methods may be required where different policy approaches are used. It is assumed here that scientists working within an asset-based policy framework would inevitably consider physical, chemical and biological processes when making environmental flows recommendations to protect or restore assets. Less obvious, but also important when working within the asset-based policy environment, is the need to consider the landscape context of the site, cumulative impacts of developments at the site and within the basin, time lags between cause and effect, and possible long-term trajectories of hydrological and ecological change (whether natural or otherwise).

The wide range of approaches to environmental flows assessment that have been developed and are in use throughout the world (Tharme, 2003; Gordon *et al.*, 2004) simply reflects the wide range of policy approaches, scientific philosophies and traditions, river types (hydrological/geomorphological/ecological characteristics), utilization of water and other river resources, and availability of technical capacity and financial resources to undertake the work. Thus, there is no single best approach that can be recommended for universal application. It is also apparent that many named 'methodologies' contain elements of other methodologies, or are a derivative of another methodology. When faced with the need to undertake an environmental flows assessment, practitioners would be better served to fashion an approach to suit their particular circumstances rather than being too concerned about faithfully following a particular documented methodology. However, there would be a major advantage in following a consistent methodology for all sites and rivers within a particular water resources jurisdiction, as higher-level water resources planning requires data that are derived in a consistent way. It may still be possible, and even desirable, to apply more than one method within a jurisdiction, provided all assessments reported a minimum set of results, using standard units and formats. For example, a rapid method could be applied inexpensively to regional assessment or sites of known low ecological value, while a more costly detailed method could be used for specific sites, or for systems known to be of high value. The methodology described in this paper is appropriate to the scale of the individual river reach, and was developed in response to a set of circumstances encountered in Taizhou, although some of these would be common in many parts of the People's Republic of China. Thus, while the approach may find general application, it is not presented as a universal solution to environmental flows assessment in the People's Republic of China.

Selection of an Appropriate Methodology

Jiaojiang Basin Case Study

The Jiaojiang Basin is located in Taizhou, a prefecture-level city in eastern Zhejiang Province (Figures 1 & 2) with a population of 5.47 million people. Zhejiang Province is undergoing a period of rapid development. In 2005, thirty of the province's cities were

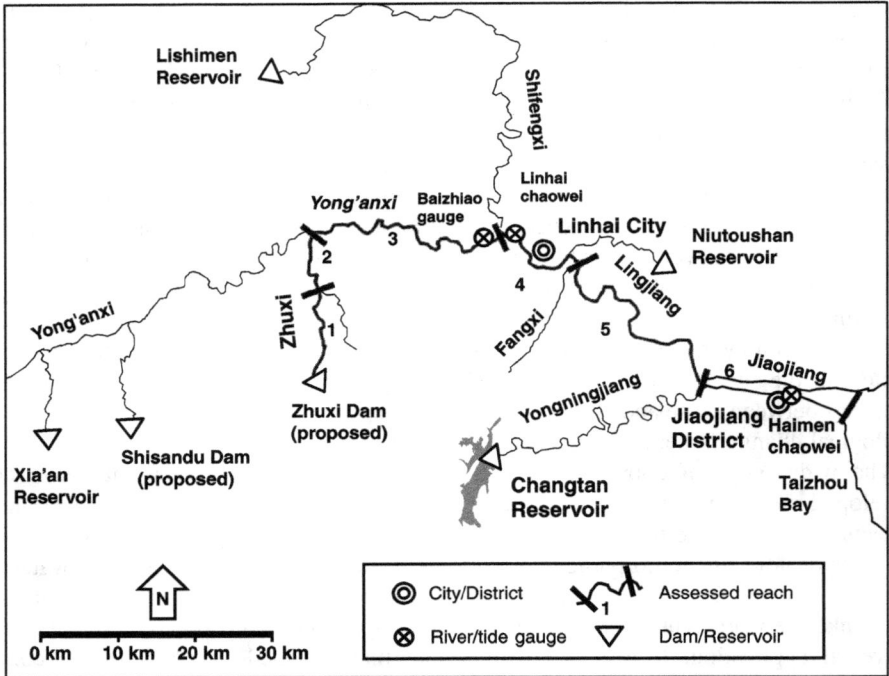

Figure 2. The six study reaches of the Jiaojiang Basin considered in the environmental flows assessment. *Note:* Not all streams and reservoirs are depicted.

listed in the All China Top 100 Counties/Cities by Comprehensive Competitiveness, published by the National Bureau of Statistics of China (Feng & Hong, 2007). This development, combined with increasing standards of living, will lead to increased demands on water supplies.

Taizhou has a humid subtropical climate and receives regular and relatively high wet season rainfall due to plum rain and typhoon influences (see Figure 3). Average annual precipitation across Taizhou is 1632 mm and average annual runoff ranges between 600 mm and 1200 mm. Floods are common in the wet season. However, a distinct dry season exists over autumn and winter. In an analysis of records from 1470 to 2005, Feng and Hong (2007) found that in Zhejiang, the 17th century was typified by frequent drought and flood, the 16th, 19th and 20th centuries were average periods, while the 18th century was a period of spasmodic drought and flood. However, in the period from 2001 to 2005, Zhejiang had one flood year (2004), two drought years and two severe drought years. In 2003 the drought lasted from summer to autumn, with the July to October precipitation being only 30% of average (Feng & Hong, 2007). Using forecasting techniques, Feng and Hong (2007) predicted that over a relatively long period into the future, Zhejiang Province will feature mostly drought years. Throughout the 20th century the drought periods were of relatively short duration, meaning that historically the region has generally had more than enough water available to meet consumptive needs. One consequence of this apparent water abundance has been the generally low awareness of the need to cater for ecological water requirements. Continued exploitation of water resources under

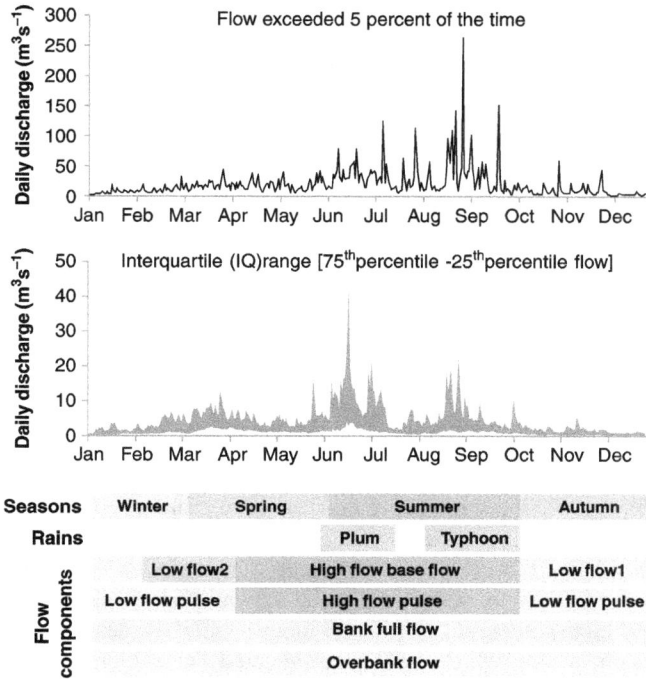

Figure 3. Distribution of some non-parametric flow statistics for each day of the year over the period 1980 to 2006 for Zhuxi downstream of proposed dam. *Note:* High flow index is flow exceeded 5 percent of the time, with range of medium flows indicated by the interquartile range. Also indicated are the seasons for climate, main rain event seasons, and environmental flows components.

conditions of more frequent drought could pose risks to the ecological health of streams in the Jiaojiang Basin.

On average, about $9.08 \times 10^9 \, \text{m}^3$ of water is available in Taizhou annually. In 2005, about $1.54 \times 10^9 \, \text{m}^3$ ($\sim 17\%$ of the total available resource) was used throughout Taizhou. The Jiaojiang Basin is the third largest in Zhejiang Province. With a total length of 208 km and catchment area of 6750 km^2, it covers 70% of Taizhou's land area. Currently, there are four major storages in the Jiaojiang Basin, plus a large number of smaller storages maintained by villages (Figure 2). Changtan Reservoir, built in the late 1950s to early 1960s on Yongningjiang is the largest, with an effective storage volume of $356 \times 10^6 \, \text{m}^3$. This storage is used predominantly for urban supply to Jiaojiang District. The other main storages are Niutoushan Reservoir, Lishimen Reservoir and Xia'an Reservoir. Two major new storages, the Zhuxi and Shisandu Reservoirs, have been proposed for tributaries of Yong'anxi in the southwest of Taizhou to accommodate projected increases in water demand. The two new storages, when combined with the Changtan Reservoir, are expected to provide sufficient supplies for development in the south of Taizhou beyond 2020. This paper is concerned only with the proposed Zhuxi Reservoir.

Zhuxi Reservoir is due to be operational by 2011. The catchment area at the dam site is 172.3 km^2 and the mean annual total flow is $192 \times 10^6 \, \text{m}^3$. The area to be inundated by the reservoir will be about 460 ha. The dam wall will be about 70 m high. The effective storage volume will be 93.5 × 10^6 m^3. The dam will also serve a flood mitigation function, and will be used for opportunistic hydropower generation. There is a general expectation that urban and industrial users will receive an annual reliability of 95 or 97%, while agricultural users will have 75% reliability (Water Entitlements and Trading Project (WET), 2007).

A method was required to assess the environmental flows needs of Zhuxi downstream of the proposed dam, plus for the affected river reaches (Yong'anxi, Lingjiang and Jiaojiang) further downstream to where the system meets Taizhou Bay. The selected method was drawn from existing methodologies, tailored to suit the local situation.

Methodological Choices

Several scientific methodologies can be applied to the problem of recommending environmental flows that will inform policy. Environmental flows methodologies were grouped by Tharme (2003) into four main categories, namely hydrological, hydraulic rating, habitat simulation (or rating) and holistic methodologies. Tharme (2003) also recognized two minor classes of methodologies: 'combination' (or hybrid) approaches which had characteristics of more than one of the four basic types, and a group termed 'other' which comprised techniques not specifically designed for environmental flows assessment but which had been adapted, or which had potential, to be used for that purpose. Methods used to determine the need for flows to maintain ecologically important geomorphological processes (sediment dynamics) could also be grouped separately from those that focus on ecological requirements per se (Gippel, 2002).

Top-down approaches assume that the entire natural flow regime is ecologically important, only allowing certain components of the flow regime to be removed from the river for off-stream consumption if it can be demonstrated that their removal presents low risk (Arthington, 1998). The top-down Benchmarking Methodology (Brizga *et al.*, 2002) undertakes basin-scale evaluation of the potential environmental impacts of future scenarios of water resource management. In this method, flow alteration data and ecological impairment data are used to set two critical benchmarks (or risk levels); the first benchmark is the limit that must be placed on flow modification to achieve protection of the natural assets of the stream or river, whereas the second benchmark represents the degree of flow alteration associated with severely degraded river conditions. These relationships are then used at the site in question to infer the likely impairment under alternative flow scenarios based on stress-condition type models. Bottom-up methodologies, such as the BBM (Building-Block Methodology) (Tharme & King, 1998), identify the key flow components that must be retained in order to maintain the ecological health of the river; theoretically, the rest of the flow can be used for other purposes, with low risk to ecosystem health. This deductive approach relies on a good conceptual understanding of the links between flows and ecological processes, and the flow recommendations relate explicitly to a set of river-specific objectives concerning the desired ecological condition.

Brown and King (2003) classified environmental flows methodologies into two categories, prescriptive and interactive, from the perspective of their usefulness as a tool for negotiating trade-offs between stakeholders. Prescriptive approaches deliver a single set of flow recommendations that best satisfy the agreed ecological objectives for managing flows in the river. If these recommendations describe the minimum set of flow components required to maintain a certain level of stream health at low risk, then it follows that failure to fully implement this regime would increase the risk of not meeting the objectives, or could mean an inevitable decline in stream health. The interactive approach formalizes this risk by presenting a range of flow scenarios that have different likely outcomes. In this way, interested parties can explicitly consider the level of risk they are prepared to take to achieve certain objectives, or they can revise their objectives as they are informed through the environmental flows assessment process. The South African DRIFT (Downstream Response to Imposed Flow Transformations) process (King *et al.*, 2003) is an example of an interactive approach.

Hydrology-only methods derive flow recommendations simply from flow records, using a flow duration statistic, or some other hydrological statistic. Most hydrological methods were initially developed using some type of empirical ecological data or theory, but ecological information is not required when applying the established methods to test sites. An example is the Tennant (Montana) Method (Tennant, 1976), which worldwide has been the most commonly applied hydrology-only methodology (Tharme, 2003), and it has seen widespread use in the People's Republic of China (Xia *et al.*, 2006; Wang, 2006; Wang *et al.*, 2007). In the Tennant Method, recommended minimum daily flows are based on percentages of the mean annual flow (MAF) (i.e. average of mean daily flow for all complete years in the record), with different percentages for winter and summer months. As the baseflow (as a percentage of MAF) is reduced, so too is the expected ecological condition of the river—from 'Excellent' (baseflow 30–50% of MAF) to 'Severely degraded' (less than 10% of MAF) (Tennant, 1976). The recommended levels are based on Tennant's observations of how stream width, depth and velocity varied with discharge on eleven streams in Montana, Wyoming and Nebraska.

The Indicators of Hydrologic Alteration (IHA) is a software programme developed by The Nature Conservancy (2007a) that calculates 33 IHA parameters and 34 Environmental Flow Component (EFC) parameters (i.e. hydrological statistics). The ecological relevance of the IHA statistics presumably derives from North American experience, although Herricks and Suen (2003) adapted it for northern Taiwan, which has a physical environment not unlike that of Taizhou. In a refinement of the original IHA method, Richter *et al.* (1997) introduced the Range of Variability Approach (RVA). Whereas the IHA identifies the degree of change in the indicators (from one period to another, or pre- and post-regulation), the RVA takes another step and develops ranges for natural variation of each characteristic. The method can be used to identify annual river management targets based on an allowable range of variation in each of the 33 IHA parameters. The method prescribes that environmental flows regime characteristics should lie within the targets for the same percentage of time as they did prior to regulation. This method has been applied in numerous environmental flows related studies in North America and has attracted interest in a few other countries (Tharme, 2003). The literature does not explicitly state how the IHA/RVA calculations are used to specify a practical flow regime that can be used to manage a river (i.e. specific recommendations for flow magnitude, frequency and duration that a dam operator or regulator could follow). The IHA User's Manual (The Nature Conservancy, 2007a) provides no guidance as to how an analysis of EFC parameters can be used to make environmental flows recommendations (other than stating that the flow components must be maintained). While IHA and EFC analysis is useful for characterizing the flow regime, a separate process is required for shaping the environmental flows recommendations. For example, Richter and Thomas (2007) discussed several environmental flows case studies in which they have been involved, but made mention of IHA and EFC only in the context of using these tools to assess the degree of flow alteration as a preliminary step in the process.

Hydraulic methods require field measurement or modelling of hydraulic parameters at the test site, and these data are related to known or assumed ecological tolerances, requirements or preferences, or geomorphological processes. The equations for modelling, and field techniques for measuring, the relationships between hydraulic parameters (depth, velocity and shear stress) and discharge are well established and can be used as a tool in any environmental flows study, regardless of the framework being followed. At the simplest level this might involve consideration of only the wetter perimeter (Gippel & Stewardson, 1998; Liu et al., 2006), while other approaches undertake a more comprehensive analysis of the hydraulics and the relationship with habitat availability. The most well known example of the latter is the Instream Flow Incremental Methodology (IFIM) (Bovee, 1982).

Holistic approaches are essentially frameworks for organizing and using flow-related data and knowledge (Brown & King, 2003). This approach is not constrained by the analytical tools, and it is not unusual to make use of several different methodologies. While the hydrological, hydraulic rating and habitat rating methods have traditionally focussed on key sport fish such as salmonids or fish with a very high conservation value, the holistic approach attempts to consider the entire ecosystem. The holistic method is usually grounded on the 'natural flow paradigm', which states that discharge variability is central to sustaining and conserving biodiversity and ecological integrity (Poff et al., 1997; Richter et al., 1997). Explicit numerical models that relate discharge to aspects of the river's geomorphology, water quality or ecology may be available, or even developed through the course of the investigation, but they are usually used as an aid to decision-making, rather than as a numerical solution to the problem of defining a suitable regulated flow regime. In many cases these models are little more than working hypotheses (Arthington & Pusey, 1993). The problem of uncertainty is usually overcome by adopting a conservative 'precautionary principle', and by recommending ongoing monitoring and adaptive management.

Arthington et al. (2006) proposed an approach to the identification of environmental flows guidelines that bridges the gap between simple hydrology-only methods and more comprehensive (but more expensive), river-specific, environmental, flow assessments. The approach is known as the Ecological Limits of Hydrologic Alteration (ELOHA) and is described in The Nature Conservancy (2007b). ELOHA is a general framework for developing scientifically-credible, environmental flows standards applicable at the multi-river (regional) scale, based on flow-ecology linkages. Development of flow-ecology response relationships for streams within regions will potentially be time consuming and costly, although some regions will already have data available on which to base such relationships. Arthington et al. (2006) recognized that in establishing flow-ecology relationships, sophisticated sampling design will be required in order to separate the effects of flow modification per se from the effects of land use that often accompany major water resource developments (Bunn & Arthington, 2002). One disadvantage of ELOHA is that it cannot be applied in regions where the streams are relatively un-impacted by flow regulation, because such areas lack opportunities to calibrate ecological response to existing gradients of flow alteration.

Most environmental flows literature is concerned with the freshwater reaches of rivers. The work that has been undertaken on estuaries (Gordon *et al.*, 2004, pp. 308–310) suggests that the methods applied to freshwater reaches can be adapted to estuaries, with the key difference being the need for hydrodynamic and salinity modelling. This is inherently more complex and expensive than modelling the hydraulics of stream flow.

Adopted Approach

In a review of methodologies potentially suitable for the People's Republic of China by WET (2007, pp. 70–110), the ELOHA approach (Arthington *et al.*, 2006) was included as part of a recommended framework for an asset-based, regional-scale assessment and implementation of environmental flows. This framework, were it to be implemented, would require considerable initial time and resources, because it would first require regionalization of hydrology and development of ecology response relationships across a range of rivers with varying degrees of flow alteration. For rivers lacking this kind of information, which is the case for the Jiaojiang Basin, Arthington *et al.* (2006) recommended application of one of the holistic methodologies. Also, The Nature Conservancy (2007b) pointed out that ELOHA does not supplant river-specific approaches for certain rivers that require more in-depth analysis. In selecting an appropriate approach to environmental flows assessment for the Jiaojiang Basin study, several factors were considered: time available, resources available, importance of river assets, length and complexity of stream under consideration, and availability of hydrological, ecological, geomorphological and hydraulic information. A comparison of these factors against the demands and limitations of various methodologies used in different parts of the world concluded that an adaptation of the river-specific, asset-based, holistic FLOWS framework (SKM *et al.*, 2002) was appropriate (WET, 2007, pp. 433–448).

The FLOWS method is a derivative of the bottom-up, prescriptive BBM (Tharme & King, 1998). This method uses a framework with a series of logical steps undertaken by a small panel of scientific experts (usually representing the fields of hydrology, hydraulics, vegetation, invertebrates, fish, geomorphology, water quality and in some cases birds) in cooperation with the river manager. There are three reporting stages and regular community consultation. Community consultation takes the form of information exchange with local water resource agency staff and interested water users from all levels, such as individual landholder, recreation groups and water supply companies. Interested stakeholders are free to observe and interact in the field inspection, supply any data they may have, query the science and make suggestions. This interaction is not about allowing the social and economic issues to influence the environmental flows process, but to provide the opportunity for the community to gain confidence in the scientific process.

The steps in the FLOWS method are: site paper (reporting phase); field inspection; issues paper (reporting phase); field topographic survey; hydraulic modelling; flows assessment workshop; flows recommendations paper (reporting phase). One fundamental difference between the BBM and FLOWS is that FLOWS incorporates seven season-specific flow components while BBM has three. The seven FLOWS components (cease to flow, low and high season baseflows, low and high season pulses, bankfull and overbank) describe the flow components typical of rivers in Victoria, Australia, which tend to be intermittent and have strong two-phase seasonality. Of course, in reality, natural flow regimes are more complex than this would suggest; simplification to a limited number of components makes practical the panel's task of assessing the required flow regime and the river operator's task of implementing the recommended regime. It is suggested here that environmental flows assessment practitioners should devise their own set of flow components based on local conditions. The broad flow components are a simplification of the typical distribution of flows throughout the year (Figure 3), but can incorporate any major well-known ecological phenomena, such as spawning runs.

The issues paper is an important document that incorporates a literature and data review, and a statement of flow objectives for ecology, geomorphology and water quality. This paper sets a vision for the river, and also separates flow and non-flow related river issues. The FLOWS method specifies professional survey of selected sites and subsequent application of a one-dimensional hydraulic model. In this way, the hydraulic parameters water depth, shear stress and velocity that are used by the scientific panel to define important ecological and geomorphological processes can be expressed as discharge, which is the unit required by managers in order to manage flows. While the hydraulic model supplies the required magnitudes of the flow components that will meet the objectives, analysis of hydrological data assists the panel to define the required flow components in terms of timing, duration, frequency and rate of rise and fall. The hydrological analysis is undertaken for modelled current and natural flow scenarios if relevant and available, or historical gauged data can be used. The panel uses these results, plus scientific literature and their collective expert judgement, to recommend the appropriate frequency and duration of flow components. For some or all of the flow components the recommended frequency and/or duration will be less than in the natural regime, as the panel is charged with recommending the minimum flow regime to meet the agreed objectives with an acceptable level of risk, not an ideal/natural regime.

Due to time and resource constraints, the bulk of the technical work on the Jiaojiang Basin environmental flows project was completed in only two weeks, while a FLOWS study would normally be undertaken over several months. This required substitution, modification or omission of certain tasks, but the critical methodological steps were retained. It is always preferable to include hydraulic assessment to support the hydrological analysis and ecological assumptions, so this aspect was retained, although in a simplified form. Staff from local authorities participated in the field inspection and hosted project meetings, and numerous interviews were held with local people involved in commercial- and subsistence-scale fishing activities. The panel members worked together exclusively on the project for two weeks, rather than gathering for a limited number of meetings spread out over a few months, as would typically happen with a FLOWS study. A significant additional step was added to the method for the Jiaojiang Basin study to allow for interactive assessment of flow options. FLOWS is a prescriptive method, whereby the panel delivers a single set of flow recommendations. These recommendations become one of the contributions to a separate environmental flows negotiation and implementation process, in which the panel members are not directly involved. For the Jiaojiang Basin study it was decided to take a scientific approach to the environmental flows trade-off process, and integrate this as a formal step in the overall method. Thus, after recommending a preferred environmental flows regime, the panel used a risk assessment methodology to provide a number of alternative flow regime options. All of these scenarios were then tested for compliance by running a hydrological model with the flow options programmed as rules. The scenarios were tested for compliance with the panel's flow recommendations and also for compliance with the expected security of supply. This interactive step is not discussed in any detail in this paper, but is described in Gippel *et al.* (2009).

Table 1. Major steps in the environmental flows assessment method used for the Jiaojiang Basin study

Methodological step	Suggested approaches to the tasks
1. Select representative reaches and sites	Use mapping, databases, reconnaissance and local knowledge to simplify river system into a manageable number of reasonably homogeneous reaches, with representative sites selected for detailed field inspection.
2. Identify ecological assets	Use existing published and unpublished literature, databases, field inspection and interviews with local river users to make a list of ecological assets.
3. Develop conceptual models	Use literature and expert knowledge to develop conceptual models that link the identified assets to key components of the flow regime and inter-related geomorphic forms and processes. The ideal model is a mathematical expression between some aspect of the asset and some aspect of the flow regime, but most will be in the form of text, or a process diagram.
4. Set management objectives for each asset and process	Objectives are expressed as the desired state of the asset, or condition of a process important to that asset. Each objective is linked to one of the flow components and is also expressed in terms of hydraulic and/or hydrological criteria that can be analysed using numerical methods. Objectives are set for flora, fauna and the geomorphic forms and processes on which they depend.
5. Hydraulic modelling and hydrological analysis	Develop hydraulic models and describe the stream hydrology to help determine the magnitude, duration, frequency and timing of flows required to meet the objectives.
6. Develop preferred environmental flows rules	Use hydraulic models, hydrological description and conceptual models to develop rules that define the minimum flow regime to sustain the ecological integrity of the river in the long term at a low level of risk.
7. Propose alternative environmental flows rules	Use a risk assessment approach to develop alternative flows rules that define flow regimes with less ambitious ecological objectives or which carry higher risk to maintenance of ecological assets in the desired state.
8. Model the water resources availability under the preferred and alternative flow options	Run a numerical water resources model with the flow rules in place to predict the degree of compliance with expected security of water supply and ecological objectives. The model is run for the flow options along with any other desired scenario, which might include proposed water resources development or climate change, for example. Assessment of the options is based on the priority given to individual objectives outlined in the initial recommendations.

The freshwater inflow requirements of the estuarine reaches of the Jiaojiang Basin study area could not be assessed using the FLOWS-derived methodology because of the lack of resources to undertake hydrodynamic modelling. An alternative risk-based approach was developed for application to these reaches. This method did not derive a flow regime specific to the estuarine reaches. Rather, it assessed the risks to the estuarine reaches associated with the regime recommended for the freshwater reach of the system under a number of water resources development scenarios.

The adopted method involved eight formal steps (Table 1). This method can be applied to situations where a water resources development is proposed or already in operation. The detail of how each task is undertaken can vary depending on data availability, resources available and the nature of the river. The eight steps refer to critical project outcomes that must be achieved in order to deliver the final product (a set of flow recommendations). Various tasks are undertaken in order to achieve these outcomes, and some of these tasks relate to more than one methodological step. For example, field inspection is not a step, but a necessary task undertaken early in the project that informs a number of steps. The following section briefly describes application of the method to the Jiaojiang Basin.

Application of the Environmental Flows Methodology

Step 1. Select Representative Reaches and Sites

The 121 km of river was divided into 6 reaches, ranging from 9 to 32 km in length. The divisions were based on hydrological boundaries (Table 2). The lower two reaches,

Table 2. The six river reaches defined for the Jiaojiang Basin study

Reach number	Stream	Description	Stream type	Boundaries	Gradient (%)	Length (km)
1	Zhuxi	Directly downstream of proposed dam	Freshwater, non-tidal	Proposed dam to first major tributary	0.242	14.2
2	Zhuxi	Downstream of first tributary inflows	Freshwater, non-tidal	First major tributary to Yong'anxi junction	0.242	9.3
3	Yong'anxi	Main stem	Freshwater, non-tidal	Zhuxi junction to Shi-fengxi junction	0.066	32.0
4	Yong'anxi	Main stem	Freshwater, tidal at non-flood times	Shifengxi junction to Fangxi junction	0.034	15.1
5	Lingjiang	Main stem estuary	Estuarine, tidal	Fangxi junction to Yogningjiang junction	0.004	31.5
6	Jiaojiang	Main stem estuary	Estuarine, tidal	Downstream of Yog-ningjiang junction to Taizhou Bay	0.000	18.8

Lingjiang and Jiaojiang, were open estuary, with the strong marine influence complicating the environmental flows assessment.

Step 2. Identify Ecological Assets

Information on ecological assets was collected from background reports, literature surveys, field visits, discussions with Bureau of Water Resources and Fisheries Bureau staff, and interviews with locals active in the fishing industry, both at commercial and subsistence scales. Overall, fish and fisheries emerged as the most highly valued asset, particularly those species of economic importance. Fish populations therefore formed a core component of the Jiaojiang Basin environmental flows assessment. However, in recognition of the strong interdependencies between fish and other components of the riverine ecosystem, such as plants (aquatic and riparian), geomorphology and water quality, these other factors were also considered for each reach.

Step 3. Develop Conceptual Models

Conceptual models grew from gaining a basic understanding of the distribution of flows in the river system. Flows are not gauged at the Zhuxi dam site, but in support of the project a time series of daily flows was generated using IQQM (Integrated Quality and Quantity Model), which is a hydrologic modelling tool developed in Australia for use in planning and evaluating water resource management policies (Department of Land and Water Conservation, 1998). A suite of hydrological statistics were produced to assist the panel, including seasonal distribution (Figure 3), flow duration, spells analysis and flood frequency analysis. Zhuxi lies within the transitional band between the inland 'Plum' and coastal 'Typhoon' hydrological zones, as mapped by Feng and Hong (2007). As a result, the annual hydrograph shows a distinct peak in medium flow indices in June (Plum rain season), and again in August–September (typhoon season), but the highest flows belong to the typhoon season (Figure 3).

The hydrological data were used in conjunction with literature review, field inspection and expert knowledge to develop conceptual models linking flow components to important physical and biological processes. The field inspection involved four days of work on the river, undertaking basic observations, such as measuring bed particle size (using Wolman pebble count, on riffles only), surveying cross-sections where possible, identifying plant species and their distributions, identifying fish species as opportunistically found, and inspecting similar nearby, but impounded, rivers. The panel also conducted interviews with people active in commercial and subsistence fishing, and held meetings with local agency staff (fisheries and water management).

Process diagrams for fish, geomorphology and vegetation were constructed as an aid to communicate the concepts to the various panel members and stakeholders. The models for each of these three main components were inter-related. For example, flows to maintain fish considered the effects of flow not just on fish directly (e.g. spawning cues), but also on flows to maintain habitat (e.g. geomorphology and macrophytes), food resources (e.g. invertebrates) and water quality (e.g. dissolved oxygen and salinity) (Figure 4). The bed of the river reaches was loose cobble to coarse gravel sized material, and on the basis of field observations and the literature, it was assumed that hyporheic flow was an important flow component. The conceptual models were used to develop the flow objectives.

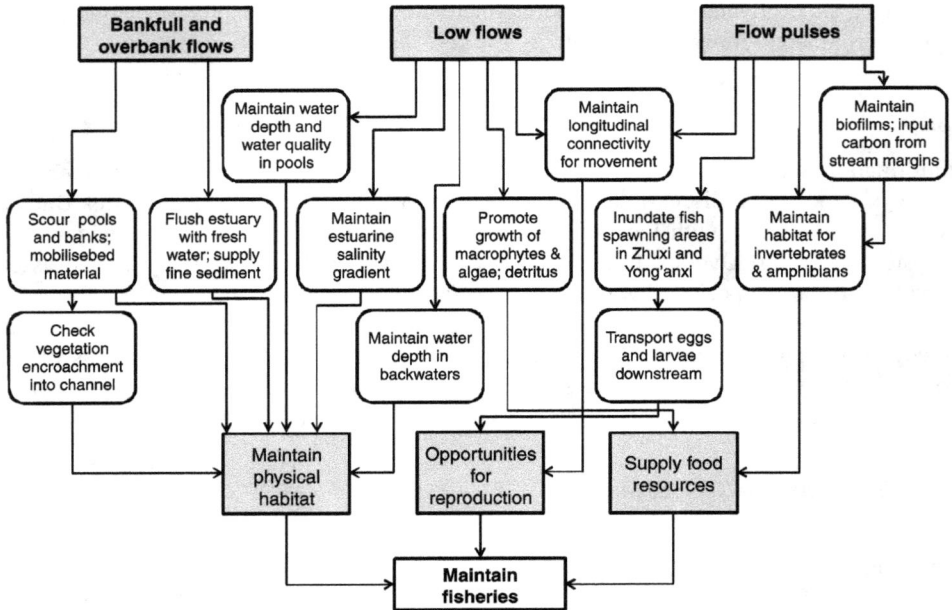

Figure 4. Conceptual process model for fish. *Note:* The model links to vegetation and geomorphology conceptual models, some elements of which are depicted here.

Step 4. Set Management Objectives for Each Asset and Process

Based on the identified environmental assets, five overarching objectives were established:

1. Maintain fish diversity and abundance.
2. Maintain subsistence and commercial fisheries.
3. Maintain water quality at a level capable of supporting objectives 1 and 2 together with existing uses of the river.
4. Maintain riparian vegetation in its current form in riparian and low-lying floodplain habitats.
5. Maintain geomorphic forms and processes that are required to support objectives 1 – 4.

Note that objective 2 contained social and economic dimensions. It was apparent from the consultation phase that fisheries were of overwhelming importance to the local community, so the objective for fish was phrased in a way that was relevant to the stakeholders. For the purposes of the environmental flows assessment, it was assumed that adequate river flows would support a diverse and abundant fish population. Of course, the sustainability of subsistence and commercial fisheries depends on many other factors not considered here.

A number of specific objectives were set for geomorphic form and process (Table 3), vegetation (Table 4) and fish (Table 5) components of the system, with each objective linked to one of the flow components (Figure 3), and also expressed in terms of hydraulic and/or hydrological criteria that allowed the objectives to be converted to a required flow magnitude. Uncertainty in the nature of linkages between each objective and hydraulic and

Table 3. Flow components relevant to geomorphic form and process objectives

ID	Objective	Flow component	Hydraulic criteria	Timing	Reach	Reference
1a	Maintain channel form	Bankfull	Morphologically defined levels	Anytime	1, 2, 3, 4	Gippel, 2001, 2002, 2005; Florsheim et al., 2008
1b	Flush fine sediment from surface of bed	Low flow pulse and High flow pulse	Critical shear stress to mobilize silts	Oct–Mar and Apr–Sep	1, 2, 3, 4	Gippel, 2001, 2002, 2005; Wu & Chou, 2004
1c	Mobilize coarse bed sediments	High flow pulse/Bankfull	Critical shear stress to mobilize >50% of bed material	Anytime	1, 2, 3, 4	Gippel, 2001, 2002, 2005
1d	Maintain key in-channel physical habitat forms (e.g. bars, benches, under-cuts)	High flow pulse/Bankfull	Morphologically defined levels	Anytime	1, 2, 3, 4	Gippel, 2001, 2002, 2005; Florsheim et al., 2008

ID = identification

Table 4. Flow components relevant to vegetation objectives

ID	Objective	Flow component	Hydraulic criteria	Timing	Reach	Reference
2a	Scour vegetation from gravel bars and transfer organic material to the stream	High flow pulse	Critical shear stress to scour aquatic plants	Apr–Sep	1, 2, 3	Groeneveld & French, 1995
2b	Maintain vegetation riparian vegetation	Overbank flow	Morphologically defined levels	Anytime	1, 2, 3, 4, 5	Naiman & Decamps (1997)
2c	Prevent encroachment from terrestrial species into river channels (e.g. Salix sp.)	Bankfull	Critical shear stress to dislodge bank vegetation	Anytime	1, 2, 3	Reid (1989), Hudson (1971)
2d	Maintain floodplain wetland communities (perennial and annual species)	Overbank flow	Morphologically defined levels	Apr–Sep	1, 2, 3, 4, 5	

Table 5. Flow components relevant to fish objectives

ID	Objective	Relevant species	Flow component	Hydraulic criteria	Timing	Reach	Reference
3a	Maintain sufficient water depth in pools for large bodied fish	Pool Guild (e.g. Eels, *Spinibarbus*, Carp species)	Low flow	D > 1.5 m in pools in reaches 1–3. D > 3 m in reaches 4 and 5	Oct–Mar	1, 2, 3, 4, 5, 6	Welcomme, 2006
3b	Maintain sufficient depth in riffles and in depositional habitats out of the main flow	Pool Guild (e.g. whitefish, catfish) and Riffle species	Low flow	D > 0.2 m	Oct–Mar	1, 2, 3, 4, 5, 6	
3c	Localized movement of resident fish	Whitefish, catfish	Low flow Pulse	D > 0.2 m over riffles	Oct–Mar	1, 2, 3	
3d	Maintenance of benthic habitats and hyporheic flushing	*Spinibarbus, Misgurnus mizolepis*	Low flow Pulse	Sufficient to flush fine sediments from gravel	Oct–Mar	1, 2, 3	
3e	Provide habitat during the high flow period, to induce spawning of grass, silver and bighead carp and to maintain transport of semi-buoyant eggs within the water column	Grass, silver, bighead carp	High flow	Mean V > 0.15–0.25 ms^{-1}	Apr–Sep	1, 2	Tang *et al.*, 1989
3f	Stimulate spawning migration and maintain longitudinal connectivity (estuary to headwaters)	Anadromous and Potamo-dromous Guilds (e.g. carp, *Coilia ectenes, Macrura reevsii*)	High flow pulse	Inundate barriers. Increase D > 0.25 m over riffles	April	1, 2, 3	Welcomme, 2006; Dudgeon, 1999
3g	Provide access to floodplain habitats	Species that spawn on floodplains (no local information supplied)	Overbank flow	Sufficient depth to inundate low lying areas of the floodplain such as wetlands	Anytime	2, 3, 4, 5, 6	Welcomme, 1985; Welcomme, 2006

ID	Objective	Guild	Flow	Non-hydraulic criteria	Season	D	Reference
3h	Flow to prevent increases in the upstream intrusion of saline water during low flows	Freshwater Estuarine Guild (e.g. *Lateolabrax japonicus*)	Low flow and High flow	Non-hydraulic criteria	All year	4, 5, 6	Welcomme, 2006; Guan et al., 2005
3i	Maintain salinity and sediment dynamics at the mouth of the estuary. Physical habitat is mud-flats and river channel	Estuarine Guild (e.g. *Mugil* spp.; *Acanthopagrus schlegelii*; purple spotted mudskipper)	Low flow	Complex hydro and sediment dynamics. Not modelled.	Oct–Mar	5, 6	Welcomme (2006), Guan et al., 1998; Guan et al., 2005
3j	Maintain salinity and sediment dynamics at the mouth of the estuary. Physical habitat is mud-flats and river channel	Estuarine Guild (e.g. *Mugil* spp.; *Acanthopagrus schlegelii*; purple spotted mudskipper)	High flow	Complex hydro- and sediment dynamics. Not modelled.	Apr–Sep	5, 6	Welcomme, 2006; Guan et al., 1998; Guan et al., 2005

Note: ID = objective identification code; D = water depth; V = velocity.

hydrologic criteria should not hinder this step, but instead be accommodated by regarding these relationships as hypotheses to be tested by subsequent studies.

Step 5. Hydraulic Modelling and Hydrological Analysis

Even when certain ecological processes can be associated with particular flow components with a high degree of confidence, it is rarely possible to make a confident direct link from an ecological process to a specific flow magnitude. Rather, such linkages are more often mediated by hydraulic factors such as water depth and velocity, and thus are more readily defined as a depth of water in the channel (e.g. a minimum depth over riffles to sustain macroinvertebrates), an elevation in the channel (e.g. corresponding to the location of a particular flow-dependent vegetation community, or the top of the banks), maximum or minimum velocities (e.g. limits of swimming capability of a fish species), or shear stresses (e.g. sufficient to mobilize bed material, or scour vegetation from the bed). Thus, the magnitude of the flow components is derived from hydraulic information. Knowledge of the channel morphology and roughness allows the hydraulic characteristics to be described for any given flow. A hydraulic model provides the numerical link between hydraulics and flow. Ideally, sites selected as representative of the defined reaches would be surveyed by professional surveyors and those data used to develop hydraulic models. Resource and time limitations prevented this approach for the Jiaojiang Basin study, so alternative hydraulic data were sought. It was beyond the resources of the project to model the freshwater inflow related hydrodynamics of the estuarine Reaches 5 and 6.

Hydraulic models were developed for Reaches 1 and 2 by surveying riffle cross-sections in the field using a laser range finder and inclinometer, and analysing the data in WinXSPRO (public domain software developed by the USDA Forest Service, Stream Systems Technology Center) (Hardy *et al.*, 2006). The hydraulic models generated rating curves of mean velocity, mean shear stress and discharge against depth. Bankfull level was defined using morphological criteria, as defined in the field. These rating curves were used to convert each of the defined geomorphological, fish and vegetation hydraulic indices into discharge values. For sediment mobilization, shear stress thresholds were computed by applying Shields Critical Shear Stress Method (Gordon *et al.*, 2004, pp. 193–195).

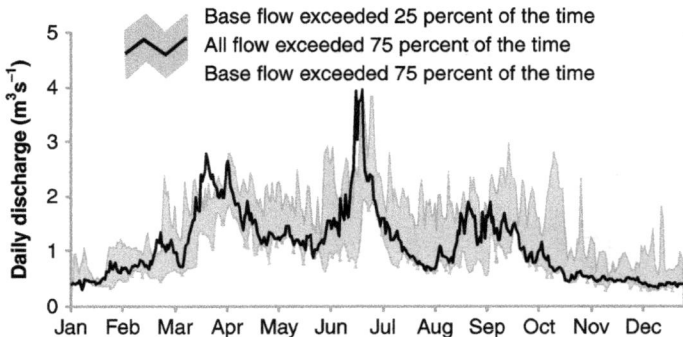

Figure 5. Distribution of low flow statistics for each day of the year over the period 1980 to 2006 for Zhuxi downstream of proposed dam. *Note:* The range of baseflows is indicated by the interquartile range of flows for times when flow was predominantly baseflow (Baseflow Index ≥ 0.9), as opposed to quickflow or transitional flow.

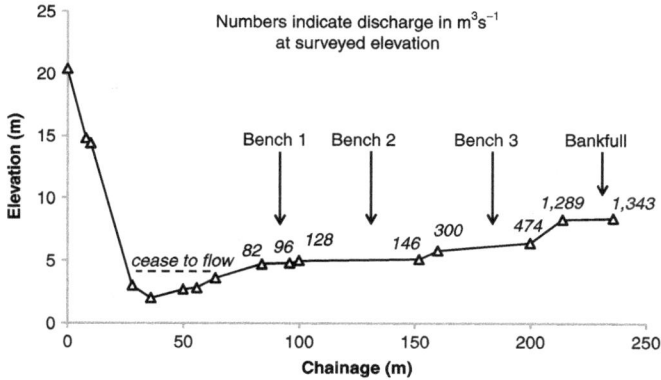

Figure 6. Cross-section at Baizhiao gauge, showing discharges corresponding to morphologically defined levels.

Unfortunately, suitable hydraulic indices for checking woody vegetation encroachment were not found in the literature, so shear stress values given in the literature for scouring grass (Reid, 1989; Hudson, 1971) were used. These are high shear stresses, and were expected to be conservative in this application.

The hydraulic models for Reaches 1 and 2 were regarded as relatively low accuracy for small discharges and shallow depths. This was because of the relatively low resolution of the survey, and the high sensitivity of the predicted flows to the roughness factor. Because of this problem the low flows and high flows (i.e. the baseflow components) were also estimated using two hydrological indices. The first index was the flow exceeded 75% of the time. The second, intuitively preferable but more complex, index was the baseflow exceeded 50% of the time. A baseflow time series was constructed by first separating baseflow and quickflow using a Lyne and Hollick (1979) recursive digital filter. The Baseflow Index (BI) is the ratio of the baseflow component of flow to total flow, such that BI $= 1$ when flow is all baseflow, and zero when all flow is stormflow. There is a transition between baseflow and quickflow when BI is between zero and 1. Periods of dominantly baseflow were separated from periods of quickflow and transitional flow on the basis that flow was strongly baseflow when BI ≥ 0.9. This was an arbitrary threshold, selected on the basis of expert judgement applied to the Jiaojiang Basin flow data. In this case, the flow exceeded 75% of the time fell between the interquartile range of baseflows (Figure 5). Thus, the two suggested indices (calculated over the defined seasons) produced similar values, and these estimates were similar to the rough estimates from the hydraulic models of the flow required to achieve the desired flow depths. For the Jiaojiang Basin study, values for baseflow components were recommended only for the low flow and high flow seasons (Figure 3), but additional variability could be introduced by specifying a baseflow for each month.

Reach 3 was not physically surveyed (due to excessive water depth). However, a gauge is located at the downstream end of this reach (Baizhiao) (Figure 2). A cross-section and rating curve were obtained for this gauge. The cross-section showed well-defined benches, and it was assumed that the top right bank corresponded to bankfull (Figure 6). At cease to flow there was depth of 1.74 m of water at the cross-section. The reason for this is unknown, but most likely it relates to a downstream control.

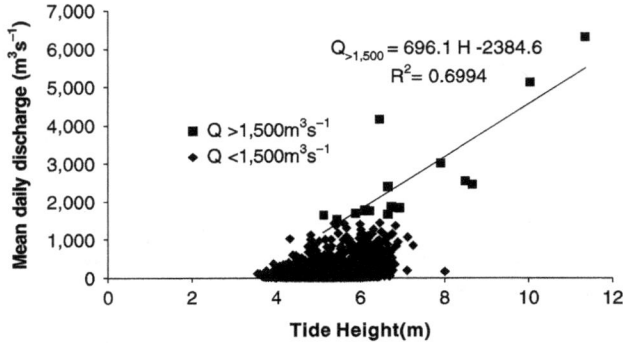

Figure 7. Relationship between daily maximum tide height at Linhai tide gauge and modelled river discharge.

Reach 4 was not physically surveyed. However, an important wetland was observed on the left bank of Yong'anxi just downstream of the Shifengxi tributary junction (Figure 2), which local knowledge suggested was inundated at a sill level of 7.5 m (relative to sea level). A tide gauge is located just downstream of the wetland (Linhai chaowei). The median tidal range at this gauge from 1996 to 2006 was 5.15 m, which is greater than the mean tidal range of 4 m reported by Guan *et al.* (2005) for the tide gauge at Haimen (now known as Jiaojiang) near the mouth of the estuary. Tidal amplification is to be expected in this estuary due to the narrowing of the channel cross-section in the upstream direction, so the entire Lingjiang reach and the lower Yong'anxi experiences a strong daily variation in water level. There has been a trend of lowering daily minimum tide levels and increasing maximum tide levels at Linhai chaowei, giving an increasing trend in daily tidal range, by around 1.8 m over the period 1996 to 2006. This is a relatively dramatic change, which could possibly be related to channel dredging, as tidal propagation in estuaries is very sensitive to water depths over the first several kilometres upstream from the estuary mouth. The apparent trend of lower minimum tide levels at Linhai has implications for the biota in Reaches 4 and 5. The slack of the low tide is now lower than it once was, exposing a greater area of the channel banks and bed, and lowering depths over the entire bed. This has contracted the area of available habitat. The trend of increasing maximum tide heights means that the upper approx. 0.5 m of the bank that was formerly mostly dry is now mostly wet, which would have altered the vegetation.

A simple hydraulic model was generated by relating the maximum tide heights to river discharge. This relationship demonstrated that peak daily water level in the river at the tide gauge was controlled by the tide for 99.6% of days. However, daily peak river height was significantly related to mean daily discharge for discharge greater than approximately $1500 \, m^3 s^{-1}$ (Figure 7). The scatter in the relationship is explained by the variable effects of tides and the fact that mean daily discharge was used rather than peak discharge (and the ratio of daily peak to daily mean discharge would have varied between events). This relationship was used to predict the mean daily discharge corresponding to the wetland sill level of 7.5 ± 0.5 m (which allowed for uncertainty in the sill level). It was assumed that on the days when the sill level was exceeded, that the peak daily flow was higher than the mean daily flow.

Environmental flows assessments are almost universally based on a time series of mean daily discharge. This is because mean daily discharge is readily available from the agencies responsible for hydrometrics, the length of the data series is manageable for analysis and plotting, and most programmes that calculate hydrological statistics require a fixed time-step, either daily or monthly. Discharge is actually measured at sub-daily time intervals, which may be fixed steps (of say 6 minutes) or variable time intervals that depend on the rate of change in stage height (i.e. less frequent measurements are made when stage height is changing slowly). When a hydraulic method is used to determine the environmental flows event magnitudes, such as flow required to inundate a bench or reach the top of bank (bankfull) these flows represent instantaneous peak discharges (Q_{PEAK}). It may not be necessary to achieve that instantaneous peak discharge for the entire day, in which case the environmental flow magnitude should be specified as both a peak (to achieve the event threshold) and a mean daily discharge (Q_{DAILY}) (as this is the common unit of discharge used by river managers, scientists and stakeholders, and it is the base unit used in hydrologic models, such as IQQM). For events that require the threshold to be achieved for a duration of one day or longer (such as pulses and baseflows), the Q_{PEAK} and Q_{DAILY} specification are the same. Although the difference between mean daily and daily instantaneous peak is well known to hydrographers and hydrologists, this would rarely be considered in an environmental flows study. Rather, it is assumed that either there is not a large difference between Q_{PEAK} and Q_{DAILY} or the hydraulic events have to be achieved for a minimum duration of one day. For the Jiaojiang Basin study neither of these assumptions were considered valid, so relationships were developed between Q_{PEAK} and Q_{DAILY}. These relationships were used for the Bankfull and Overbank flow components; the flow pulses were all recommended to be delivered for a duration of one day or longer, so these thresholds were specified as mean daily discharge.

Design flood discharges were supplied for the Zhuxi Dam site and for a location close to the end of Zhuxi. For the IQQM-modelled daily flow series for Zhuxi, discharges corresponding to a range of average recurrence intervals from 1–20 years were calculated. These two sets of data allowed a comparison to be made between peak instantaneous discharge and mean daily discharge for flood events for Zhuxi at the dam site, and at the end of Zhuxi. This comparison resulted in highly significant general relationships for converting Q_{PEAK} (the hydraulically defined threshold) to Q_{DAILY} (for use in IQQM modelling):

$$\text{Zhuxi at Dam site (Reach 1)} : Q_{DAILY} = 0.2922\,Q_{PEAK} + 23.7$$

$$\text{Zhuxi at end of creek (Reach 2)} : Q_{DAILY} = 0.3445\,Q_{PEAK} + 48.8$$

One year of sub-daily flow observations at Baizhiao gauge on Yong'anxi upstream of Shifengxi junction (Reach 3) were supplied (for 1997). From these data a highly significant relationship was established between mean daily discharge and daily peak instantaneous discharge:

$$\text{Yong'anxi at Baizhiao (Reach 3)} : Q_{DAILY} = 0.7334\,Q_{PEAK} + 3.1$$

No data were available to model the relationship between peak and mean daily discharge for Reach 4. For this reach, it was assumed that the peak was a little closer to the mean compared to Reach 3, so a factor of 1.25 (i.e. inverse of 0.8) was used to convert mean daily discharge to instantaneous discharge thresholds required for wetland inundation.

Step 6. Develop Preferred Environmental Flows Rules

Baseflow magnitudes were determined from a hydrological index. The panel did not recommend supplying a constant minimum baseflow at all times, although this would be technically possible by draining water from the storage as necessary. Rather, the baseflows were specified as the recommended value or natural flow (i.e. without the water resources development), whichever is lower, which would introduce variability in the baseflows. This approach to implementation requires availability of a model that predicts the natural flows at each compliance point in real time. In order to fully specify the event components (Pulses, Bankfull and Overbank), it was necessary to describe them in terms of magnitude, frequency, duration, timing and rates of change. Event timing was specified as within a pre-determined season (according to natural seasonality), while event magnitudes were derived from hydraulic relationships. Event frequencies and durations can sometimes be specified on the basis of known flow-ecology relationships, but at best these would be available for only a small proportion of the ecological assets being considered. The alternative approach, which was used in the Jiaojiang Basin study, is to set the frequencies and durations of events on the basis of hydrological statistics calculated from the modelled flow time series, although this is not a trivial task. Separate examination of flood frequency curves, flow duration curves and monthly or seasonal flow distribution data (which typically would be prepared by the panel's hydrologist) invariably leads to over-estimation of the frequency of flow pulses in the flow series, because for pulses to be counted they have to mutually satisfy the magnitude, timing, duration and frequency requirements. One sure way for the expert panel to lose credibility in the eyes of water users and managers would be to carelessly recommend events of a certain generous duration and frequency, only to find out later that such events rarely or never occurred in the natural flow series. To our knowledge there is no software available commercially or in the public domain that will conveniently perform the required analysis (an interactive tool is required). Ordinary spells analysis lacks the required sophistication. For the Jiaojiang Basin study, a special interactive compliance checking programme was used to undertake this analysis.

Once the magnitudes of events were established, it was possible to determine the maximum allowable and target rates of rise and fall. Rates of rise and fall were calculated for the Pulses, Bankfull and Overbank events for each relevant reach on the basis of the patterns in the modelled flow series. First, the components were separated from the flow time series. This was done, for each component, by selecting all flows equal to the magnitude of the flow component $\pm 20\%$ (the results were relatively insensitive to the width of this band). Then, for each flow component the rises and falls associated with all the events identified in the time series were separated, and descriptive statistics calculated for each. The maximum allowable rate of rise and fall was described by the rate exceeded 5% of the time. The median values were calculated as an index of the normal target rates of rise and fall. The target rate of rise was two days for most of the Pulses. The target rise for Bankfull and Overbank remained at one day except in the case of Bankfull for Reach 3. In general, rates of fall were lower than rates of rise, but they were still relatively fast. Maximum allowable rates of fall for Pulses were 1–2 days, and one day for Bankfull and Overbank. Target rates of fall for Pulses were 2–3 days, and 1–2 days for Bankfull and Overbank. This analysis highlighted the very flashy nature of the pulses and flood events in streams of the Jiaojiang Basin.

Table 6. Summary of flow requirements for Reach 1, immediately downstream of proposed dam on Zhuxi

Flow component	Timing	Months	Q_{PEAK} ($m^3 s^{-1}$)	Q_{DAILY} ($m^3 s^{-1}$)	Frequency (per year)	Duration	Rise and fall target (max.) ($m^3 s^{-1}$)	Flow objective ID
Low flow	Low flow season	Oct–Mar	0.5	0.5	Continuous or less than if natural	Not relevant	Not relevant	3a, 3b
High flow	High flow season	April–Sep	1.3	1.3	Continuous or less than if natural	Not relevant	Not relevant	3e
Low flow pulse	Low flow season	Oct–Dec; Jan–Mar	20	20	2	1 day in Oct–Dec and 1 day in Jan–Mar	Rise: + 13(+20) Fall: − 10(−14)	1b 2a 3c, 3d
High flow pulse	Spawning period	April	20	20	1	1 day	Rise: + 13(+20) Fall: − 10(−14)	1b 2a 3d, 3f
High flow pulse	High flow season	May–Sep	20	20	4	2 consecutive days	Rise: + 13(+20) Fall: − 10(−14)	3f
Bankfull	High flow season	Anytime	524	177	0.52	1 day; achieve Q_{PEAK} and Q_{DAILY} on day of peak	Rise: + 150(+190) Fall: − 134(−165)	1a, 1c, 1d 2c
Overbank	High flow season	Anytime	571	191	0.52	1 day; achieve Q_{PEAK} and Q_{DAILY} on day of peak	Rise: + 158(+200) Fall: − 141(−182)	2b, 2d

Note: Frequency for Bankfull and Overbank is probability of exceedance calculated from the partial duration series.

Table 7. Summary of flow requirements for Reach 4, Yong'anxi from Shifengxi junction through Linhai to salt wedge limit at Fangxi junction

Flow component	Timing	Months	Q_{PEAK} ($m^3 s^{-1}$)	Q_{DAILY} $m^3 s^{-1}$	Frequency (per year)	Duration	Rise/fall max. (target) ($m^3 s^{-1}$)	Comment	Flow objective ID
Low flow	Low flow season 1	Oct–Jan	16	16	Continuous or less than if natural	Not relevant	Not relevant		3h, 3i
Low flow	Low flow season 2	Feb–Mar	32	32	Continuous or less than if natural	Not relevant	Not relevant	Will maintain low salinity flow on the ebb tide in the upper part of the reach	3h, 3i
High flow	High flow season	April–Sep	46	46	Continuous or less than if natural	Not relevant	Not relevant		3h, 3j
Low flow pulse / High flow pulse					None specified. Pulses will not influence water level. A minor and temporary influence on downstream salinity gradient is possible, but cannot be linked to known ecological processes				
Bankfull[†]	Anytime	Anytime	3545	2836	0.48	1 day; achieve Q_{PEAK} and Q_{DAILY} on day of peak	Rise: $+ 3,674 (+5,171)$ Fall: $- 2,963 (-4,481)$	Event to reach 7.5 m at wetland	1a, 1c, 1d 2b, 2d 3g, 3i, 3j
Overbank	Not specified								

Note: Frequency for Bankfull is probability of exceedance calculated from the partial duration series.

To illustrate the way environmental flows recommendations were specified (flow rules), the preferred regimes are provided here for Reach 1 (Table 6) and Reach 4 (Table 7). The flows recommended for the other reaches are detailed in WET (2007). Note that each flow component was specified in sufficient detail that a river manager would be able to implement it without any ambiguity. In systems under high water demand, it is possible that the recommendations will be challenged by stakeholders, which is why it is important that the recommendations are presented together with the direct links back to the original agreed objectives.

Hyporheic flow, the movement of water through the subsurface (Poole *et al.*, 2006), was observed in the field, but little is known of the typical flow rates or the ecological importance of hyporheic flow in this setting. Habitat patch diversity in the hyporheic zone should increase with variation in geomorphologic structure and discharge regime because changes in patterns of groundwater movement are driven by interactions between channel morphology and hydrology. Thus, it is important to maintain hyporheic flows in this river system. The compliance points for the environmental flows recommended for the Zhuxi study were located at the downstream end of the defined reaches, which effectively means that the hyporheic component must be supplied in order that flows will manifest at the surface. To comply with the surface flow requirement, it may be necessary to release more than the recommended threshold discharge in order to first supply the hyporheic flow component. This will require adaptive management.

Step 7. Propose Alternative Environmental Flows Rules, and Step 8. Model the Water Resources Availability

Steps 7 and 8 were the interactive component of the environmental flows assessment methodology. These steps involved a risk assessment approach to development of alternative flow recommendations and IQQM modelling to ascertain the impact of various flow scenarios on security of supply and in-stream flows. These steps are the subject of a separate paper (Gippel *et al.*, 2009).

Assessment of Freshwater Flows for the Estuarine Reaches

It was not possible to apply the methodology used for the freshwater reaches to the estuarine Reaches 5 and 6. As an alternative, a risk assessment was undertaken to explore the potential impacts associated with five IQQM-modelled future scenarios (involving dams on Zhuxi and Yong'anxi). Three key hydrological change indices were assessed: reduction in summer baseflow; reduction in winter baseflow; and reduction in the frequency of bankfull floods. While the open Jiaojiang estuary is dominated by marine influences, freshwater flows are also ecologically important. During high river flow events the saline estuarine water is pushed out to sea. This causes a major disruption to the normal ebb and flow cycle; the biota have evolved to either take advantage of this sudden and perhaps persistent change in the hydraulics and salinity regime, or to minimize negative impacts. A significant reduction in the frequency of major runoff events, or similarly, a significant reduction in the volume of water passing to the estuary during major rainfall events, could have significant ecological consequences. Baseflows determine the longitudinal position of the salinity gradient during non-flood times. Weaker baseflows allow saline water to penetrate further upstream.

For each of the three hydrological change indices, a list was made of the ecological assets that were potentially at risk of impairment. The assets were rated according to three conservation status classes, with the consequence of change (consequence) being higher the higher the conservation status. Degree of change (likelihood of impairment due to hydrological change) was ranked into four classes based on the calculated hydrological statistics. The product of consequence and likelihood gave risk of impairment, which was grouped into five classes, ranging from insignificant to very high.

The five water resources development scenarios examined appeared to pose a fairly minimal threat to the ecological assets in Reaches 5 and 6, with the greatest, yet still only moderate, threat being posed to the upstream movement of migratory fish and to estuarine resident fish species. The main threat to the latter group would likely be a change in the distribution and availability of suitable spawning habitats, which are often located in areas with salinity levels below that of seawater. Reductions in High flow baseflows during the spawning period may cause changes in salinities in the middle and upper estuary that could isolate appropriate spawning sites, in terms of habitat, from areas with appropriate water chemistry. Overall, it would appear that construction of dams on Zhuxi and Yong'anxi would cause small but incremental changes in the baseflow volume and flood magnitude and frequency in Reaches 5 and 6. This incremental change highlights the importance of assessing the impacts of water resource development at the basin scale, as further development in other sub-catchments would almost certainly increase the risk posed to ecological assets over the levels revealed by this analysis.

Discussion and Conclusion

The flow recommendations made here using the asset-based holistic approach (incorporating both hydraulic and hydrologic tools) for the Jiaojiang Basin were compared with the recommendations made by the Tennant Method (Tennant, 1976). The only aspects of Tennant Method recommendations that bore any resemblance to the recommendations made here were the basefow components (Low flows and High flows). For these, the Tennant management condition class that compared closest to the recommendations made here varied from 'Poor' at Reach 1 downstream of Zhuxi Dam to 'Excellent' at Reach 4 Yong'anxi near the wetland. The lack of correspondence revealed in this comparison should serve as a warning for anyone contemplating using the Tennant Method outside of the region where it was initially developed, especially in high gradient streams (see also Mann, 2006).

The magnitudes, frequencies and durations of the recommended flow components for the Jiaojiang Basin stream reaches were compared with the standard hydrological statistics calculated by the IHA (Indicators of Hydrological Alteration) (The Nature Conservancy, 2007a). To summarize, there was no set of IHA parameters that consistently predicted the recommended flow components, although a few statistics did give similar values for some flow components in some reaches. The lack of a consistent comparison between any of the IHA statistics and the recommended flow components was not unexpected. First, the Jiaojiang Basin study used carefully selected seasons to match the local hydrological and ecological situation. Secondly, this study selected event magnitudes on the basis of hydraulic thresholds, and because the river's geomorphologic characteristics naturally varied downstream, a consistent relationship between hydrological indices and relative water levels would not be expected. The IHA statistics could have been forced to predict the recommended flow magnitudes by adjusting the flow thresholds in the computer programme, but the relevant point is that the information necessary to do this would only become available after hydraulic analysis has been undertaken. The failure of hydrology-only methodologies to take into account the downstream change in the relationship between a river's geomorphic and hydrologic characteristics (i.e. expressed as hydraulics) is a major weakness that cannot be easily overcome. Most hydrology-only methods use flow-ecology relationships that were developed from experience in certain systems, but they can be presented as generic (e.g. IHA), or often interpreted that way (e.g. Tennant Method). It is more appropriate to use flow-ecology relationships specific to the river under investigation.

A number of non-flow related issues were identified as impacting the streams of the Jiaojiang Basin: sand and gravel extraction, estuary channel dredging, pollution and over-fishing. These represent threats to the success of any implemented environmental flows regime in achieving the objective of a healthy river. Another important issue has to do with catchment planning. It is possible to both recommend and implement an environmental flows regime to protect the ecological assets of the Jiaojiang Basin from the Zhuxi dam development. However, as further development expands across the basin, maintaining ecological assets, particularly in the lower reaches of the river system, will become a much greater challenge, both in terms of water availability, and in ensuring that diversions and releases are managed in a coordinated fashion. At the simplest level, the setting of a cap (stated in water volume) on basin-wide water use will provide the appropriate starting point from which to ensure water can be managed in a sustainable fashion. Without this initial constraint to ensure some water is available to protect environment assets, the development of environmental flows regimes could become little more than an academic exercise.

The asset-based holistic environmental flows methodology proposed here and applied to the Jiaojiang Basin downstream of the proposed Zhuxi dam was based around a number of existing methodologies outlined in the literature. The methodology broadly combines information on ecological and other assets associated with the river system (in this case fish, vegetation, water quality and geomorphology) together with information that links these assets to aspects of the flow regime via hydraulic relationships. This is a site-based methodology, and it requires a medium-level effort and budget. The methodology hinges on being able to gain a basic understanding of the ecology and geomorphology of the stream system, having daily flow series available and having the capacity to develop hydraulic models. It was found here that even relatively simple hydraulic models based on rapid field survey and existing relationships produced highly valuable information that allowed development of flow regimes that were specific to the needs of the ecological assets identified at the site.

Acknowledgements

This paper is the result of a project undertaken under the auspices of the Australian Department of the Environment, Water, Heritage and the Arts and the Chinese Ministry of Water Resources, with funding provided by AusAID, the Australian Agency for International Development. The Department of Water Resources, Zhejiang Province, supported the field component.

References

Arthington, A. H. (1998) *Comparative Evaluation of Environmental Flow Assessment Techniques: Review of Holistic Methodologies* LWRRDC Occasional Paper No. 26/98 (Canberra: Land and Water Resources Research & Development Corporation).

Arthington, A. H. & Pusey, B. J. (1993) In-stream flow management in Australia: Methods, deficiencies and future directions, *Australian Biology*, 6, pp. 52–60.

Arthington, A. H., Bunn, S. E., Poff, N. L. & Naiman, R. J. (2006) The challenge of providing environmental flow rules to sustain river ecosystems, *Ecological Applications*, 16, pp. 1311–1318.

Bovee, K. D. (1982) *A Guide to Stream Habitat Assessment Using the Instream Flow Incremental Methodology* Instream Flow Information Paper 12, FWS/OBS-82/26 (Fort Collins, CO: Cooperative Instream Flow Services Group, U.S. Fish and Wildlife Service).

Brizga, S. O., Arthington, A. H., Choy, S. C., Kenard, M. J., Mackay, S. J., Pusey, B. J. & Werren, G. L. (2002) Benchmarking, a 'Top Down'' methodology for assessing environmental flow needs in Australian rivers, in: *Proceedings of the 4th Ecohydraulics Conference*, March 4–8, Cape Town, South Africa (CD-ROM).

Brown, C. A. & King, J. M. (2003) Environmental flows: concepts and methods, *Water Resources and Environment Technical Note C1*, R. Davis & R. Hirji (Eds) (Washington, DC: The World Bank).

Bunn, S. E. & Arthington, A. H. (2002) Basic principles and ecological consequences of altered flow regimes for aquatic biodiversity, *Environmental Management*, 30, pp. 492–507.

Department of Land and Water Conservation (1998) *Integrated Quantity Quality Model (IQQM) User Manual*, Version 6.33 (Parramatta, NSW: New South Wales Government).

Dudgeon, D. (1999) *Tropical Asian Streams: Zoobenthos, Ecology and Conservation* (Hong Kong: Hong Kong University Press).

Feng, L. & Hong, W. (2007) Characteristics of drought and flood in Zhejiang Province, East China: Past and future, *Chinese Geographical Science*, 17(3), pp. 257–264.

Florsheim, J. L., Mount, J. F. & Chin, A. (2008) Bank erosion as a desirable attribute of rivers, *BioScience*, 56(6), pp. 519–529.

Gippel, C. J. (2001) Hydrological analyses for environmental flow assessment, in: F. Ghassemi & P. Whetton (Eds) *Proceedings MODSIM 2001*, International Congress on Modeling and Simulation, The Australian National University, Canberra, pp. 873–880 (Canberra: Modeling & Simulation Society of Australia & New Zealand).

Gippel, C. J. (2002) Geomorphic issues associated with environmental flow assessment in alluvial non-tidal rivers, *Australian Journal of Water Resources*, 5(1), pp. 3–19.

Gippel, C. J. (2005) Environmental flows: Managing hydrological environments, in: M. Anderson (Ed.) *Encyclopaedia of Hydrological Sciences*, pp. 2953–2972 (Chichester: John Wiley & Sons, Ltd.).

Gippel, C. J. & Stewardson, M. J. (1998) Use of wetted perimeter in defining minimum environmental flows, *Regulated Rivers Research & Management*, 14(1), pp. 53–67.

Gippel, C. J., Cosier, M., Markar, S. & Liu, C. (2009) Balancing environmental flows needs and water supply reliability, *International Journal of Water Resources Development*, 25(2), pp. 331–354.

Gordon, N. D., McMahon, T. A., Finlayson, B. L., Gippel, C. J. & Nathan, R. J. (2004) *Stream Hydrology: An Introduction for Ecologists*, 2nd ed (Chichester: John Wiley & Sons, Ltd.).

Groeneveld, D. P. & French, R. H. (1995) Hydrodynamic control of an emergent aquatic plant (*Scirpus acutus*) in open channels, *Water Resources Bulletin*, 31, pp. 505–514.

Guan, W. B., Wolanski, E. & Dong, L. X. (1998) Cohesive sediment transport in the Jiaojiang River Estuary, China, *Estuarine, Coastal and Shelf Science*, 46, pp. 861–871.

Guan, W. B., Kot, S. C. & Wolanski, E. (2005) 3-D fluid-mud dynamics in the Jiaojiang Estuary, China, *Estuarine, Coastal and Shelf Science*, 65, pp. 747–762.

Hardy, T., Panja, P. & Mathias, D. (2005) *WinXSPRO: A Channel Cross-Section Analyzer User's Manual Version 3.2. Gen. Tech. Rep. RMRS-GTR-147* (Fort Collins, CO: U.S. Department of Agriculture, Forest Service, Rocky Mountain Research Station).

Herricks, E. E. & Suen, J. P. (2003) Ecological design in Taiwan's rivers: performance expectations considering hydrologic variability, *International Workshop on Ecohydraulics and Eco-rivers Engineering*, Taipei, Taiwan, 2–3 October 2003, pp. 25–40. Available at: http://www.wra.gov.tw/public/Attachment/41110465971.pdf (accessed 1st December 2007).

Hudson, N. (1971) *Soil Conservation* (Ithaca: Cornell University Press).

Jiang, D., Wang, H. & Li, L. (2006) Progress in ecological and environmental water requirements research and applications in China, *Water International*, 31(2), pp. 145–156.

Jiang, D. (2007) Progress in environmental flows research and applications in China. Paper presented at the 10th International River Symposium and Environmental Flows Conference, Brisbane, September. Available at http://www.riversymposium.com.

King, J. M., Brown, C. A. & Sabet, H. (2003) A scenario-based holistic approach to environmental flow assessments for rivers, *River Research and Applications*, 19(5–6), pp. 619–639.

Liu, S., Mo, X., Xia, J., Liu, C., Lin, Z., Men, B. & Ji, L. (2006) Estimating the minimum in-stream flow requirements via wetted perimeter method based on curvature and slope techniques, *Journal of Geographical Sciences*, 16(2), pp. 242–250.

Lyne, V. & Hollick, M. (1979) Stochastic time-variable rainfall-runoff modeling, in: *Hydrology and Water Resources Symposium, Perth: Institution of Engineers Australia National Conference Publication 79/10*, pp. 89–93 (Canberra: Institution of Engineers Australia).

Mann, J. L. (2006) Instream flow methodologies: An evaluation of the Tennant Method for higher gradient streams in the National Forest System lands in the western U.S., Master of Science thesis, Colorado State University.

Naiman, R. J. & Decamps, H. (1997) The ecology of interfaces: Riparian zones, *Annual Review of Ecology and Systematics*, 28, pp. 621–658.

Poff, N. L., Allan, J. D., Bain, M. B., Karr, J. R., Prestegaard, K. L., Richter, B. D., Sparks, R. E. & Stromberg, J. C. (1997) The natural flow regime, a paradigm for river conservation and restoration, *BioScience*, 47, pp. 769–784.

Poole, G. C., Stanford, J. A., Running, S. W. & Frissell, J. A. (2006) Multiscale geomorphic drivers of groundwater flow paths: Subsurface hydrologic dynamics and hyporheic habitat diversity, *Journal North American Benthological Society*, 25(2), pp. 288–303.

Reid, L. M. (1989). Channel incision by surface runoff in grassland catchments, PhD dissertation, University of Washington.

Richter, B. D., Baumgartner, J. V., Wigington, R. & Braun, D. P. (1997) How much water does a river need? *Freshwater Biology*, 37, pp. 231–249.

Richter, B. D. & Thomas, G. A. (2007) Restoring environmental flows by modifying dam operations, *Ecology and Society*, 12(1), 12. Available at http://www.ecologyandsociety.org/vol12/iss1/art12/ (accessed 24 September 2008).

SKM, CRC for Freshwater Ecology, Freshwater Ecology (NRE) & Lloyd Environmental Consultants (2002) *The FLOWS Method: A Method for Determining Environmental Water Requirements in Victoria*. Report prepared by Sinclair Knight Merz, the Cooperative Research Centre for Freshwater Ecology, Freshwater Ecology (NRE) and Lloyd Environmental Consultants (Melbourne: Department of Natural Resources and Environment). Available at http://www.envict.org.au/file/Flows_Methodology.pdf (accessed 24 September 2008).

Tang, M., Huang, D., Huang, L., Xiang, F. & Yin, W. (1989) An evaluation of the hydraulic characteristics of the eggs of grass carp, black carp, silver carp and bighead carp and a preliminary prediction of the incubation conditions for them in the proposed Three Gorges reservoir, *Reservoir Fisheries*, 4, pp. 26–30.

Tennant, D. L. (1976) Instream flow regimens for fish, wildlife, recreation and related environmental resources, *Fisheries*, 1(4), pp. 6–10.

Tharme, R. E. (2003) A global perspective on environmental flow assessment: Emerging trends in the development and application of environmental flow methodologies for rivers, *River Research and Applications*, 19(5–6), pp. 397–441.

Tharme, R. E. & King, J. M. (1998) *Development of the Building Block Methodology for Instream Flow Assessments and Supporting Research on the Effects of Different Magnitude Flows on Riverine Ecosystems*. WRC Report no. 576/1/98 (Cape Town: Freshwater Research Unit, University of Cape Town).

The Nature Conservancy (2007a) *Indicators of Hydrologic Alteration Version 7 User's Manual* (Arlington, VA: The Nature Conservancy). Available at http://www.nature.org/initiatives/freshwater/conservationtools/art17004.html (accessed 24 September 2008).

The Nature Conservancy (2007b) *Ecological Limits of Hydrologic Alteration Integrating Environmental Flows with regional water management* (Arlington, VA: The Nature Conservancy). Available at http://www.nature.org/initiatives/freshwater/files/eloha_final_single_page_low_res.pdf (accessed 24 September 2008).

Wang, X. (2006) *Research on Ecological Water Requirements of China's River Courses*, Australia-China Environment Development Program, Water Entitlements and Trading Project. Australian Department of Environment, Water, Heritage and the Arts, and the Chinese Ministry of Water Resources, with funding provided by the Australian Agency for International Development, September.

Wang, X., Zang, Y. & Liu, C. (2007) Water quantity-quality combined evaluation method for rivers' water requirements of the instream enviroment in dualistic water cyle. A case study of Liaohe River Basin, *Journal of Geographical Sciences*, 17(3), pp. 304–316.

WET (2007) *Water Entitlements and Trading Project (WET Phase 2) Final Report* December 2007 [in English and Chinese] (Beijing: Ministry of Water Resources, People's Republic of China and Canberra: Department of the Environment, Water, Heritage and the Arts, Australian Government). Available at: http://www.environment.gov.au/water/action/international/wet2.html

Welcomme, R. L. (1985) *River Fisheries* FAO Technical Bulletin 262 (Rome: The Food and Agriculture Organization of the United Nations).

Welcomme, R. L., Winemiller, K. O. & Cowx, I. G. (2006) Fish environmental guilds as a tool for assessment of ecological condition of rivers, *River Research and Applications*, 22, pp. 377–396.

Wu, F. C. & Chou, Y. J. (2004) Tradeoffs associated with sediment-maintenance flushing flows: A simulation approach to exploring non-inferior options, *River Research and Applications*, 20, pp. 591–604.

Xia, J., Feng, H-L., Zhan, C-S. & Niu, C-W. (2006) Determination of a reasonable percentage for ecological water-use in the Haihe River Basin, China, *Pedosphere*, 16(1), pp. 33–42.

Balancing Environmental Flows Needs and Water Supply Reliability

CHRISTOPHER J. GIPPEL, MARTIN COSIER, SHARMIL MARKAR &
CHANGSHUN LIU

ABSTRACT *This paper describes an approach to the integration of environmental flows recommendations into water resources planning, and demonstrates its application to a case study in the Jiaojiang Basin, Taizhou, Zhejiang Province, the People's Republic of China. In this approach, environmental flows recommendations were provided to the process as a preferred regime, and also as one or more sub-optimal regimes. A risk assessment approach was used to derive the sub-optimal regimes from the preferred regime. The environmental flows rules were then incorporated into a wider water resources model which allowed testing of any number of development scenarios. The model-predicted daily time series' of river flows were passed through a sophisticated form of spells analysis to evaluate the degree of compliance with a specified environmental flows regime. This degree of compliance was balanced against the predicted security of supply to water users. This integrated approach allowed for a greater appreciation of environmental concerns by planners. It also provided an opportunity to the scientists who undertook the environmental flows assessment to contribute to the process of making rational trade-offs between risks to the environment and gains in security of supply.*

Introduction

Environmental flow methodologies can be either prescriptive or interactive (Brown & King, 2003). Prescriptive approaches deliver a single set of flow recommendations that best satisfy the agreed ecological objectives for managing flows in the river. If these recommendations describe the minimum set of flow components required to maintain a certain level of stream health at low risk, then it follows that failure to fully implement this regime would increase the risk of not meeting the objectives, or could mean an inevitable decline in stream health. The interactive approach formalizes this risk by presenting a range of flow scenarios that have different likely outcomes (e.g. King *et al.*, 2003). In this way, interested parties can explicitly consider the level of risk they are prepared to take to achieve certain objectives, or they can revise their objectives as they are informed through the environmental flows assessment process. An interactive approach also recognizes the iterative nature of the supply and demand side of the water resources allocation planning process. With often competing needs for different consumptive purposes, the planning process needs to be able to balance the needs of different users throughout the basin, considering the implications of different possible approaches. This should happen with an understanding of the environmental implications of the various options.

Gippel *et al.* (2009) have described a site-based, ecological asset-based, holistic environmental flows assessment approach, and demonstrated application of six of the eight steps suggested under this approach to the Jiaojiang (Jiao River) Basin in Taizhou, Zhejiang Province, People's Republic of China. This paper describes and illustrates the final two interactive steps of the methodology. These interactive steps involve integration of the environmental flows assessment procedure into the wider water resources planning process, which is concerned with the reliable supply of water to consumptive users as well as providing for the ecological needs of rivers.

The process of planning for major water resources developments involves a number of modelling stages before final designs and operating rules are agreed. The review by Vogel *et al.* (2007) found seven published papers that have addressed the problem of optimization of reservoir operation to balance water supply reliability and ecological needs, with the two most recent papers being by Harman and Stewardson (2005) and Suen & Eheart (2006). Prior to optimization of reservoir release strategies it is first necessary to be able to simulate reservoir operations. Simulation models are used to evaluate the fundamentals, rather than the details, of the trade-offs involved between environmental and consumptive water requirements. This paper is concerned only with the simulation stage of the planning process.

The first published study that modelled the trade-off between security of supply for consumptive purposes and flows for environmental benefit was by Palmer and Snyder (1985). They defined the environmental flow requirements as simple minimum discharges for each month (i.e. not a multi-component regime), and alternative flows scenarios were generated by factoring of these monthly values (i.e. there was only a loose, assumed relationship between the flows scenarios and the risk to the environment). Gippel and Stewardson (1995) evaluated the impacts of various environmental flows regimes (specified as minimum discharges for each month) and system operating strategies on the date when the next resource augmentation would be required. Wollmuth and Eheart (2000) evaluated the impact on environmental flows resulting from five different release rules for reservoir systems under irrigation demands. Cai and Rosegrant (2004) undertook a scenario analysis of some water development strategies in the Yellow River Basin, People's Republic of China. They evaluated the trade-off between agriculture water demand and ecological water demand, but their modelling scenarios were constrained by using a fixed ecological water demand. Vogel *et al.* (2007) constructed relationships between storage, yield and environmental flows using a single, overall measure of environmental flow protection. They recognized that a single metric did not adequately quantify environmental flow protection, but nonetheless introduced a single metric, called the ecodeficit, which was based on a flow duration curve. That study, like several others included in the review by Vogel *et al.* (2007), tacitly assumed that the benefits of environmental flows to the river were, like the benefits of water for consumptive use, a continuous function of the volume of water provided. This follows from the 'natural flow paradigm' (Poff *et al.*, 1997; Richter *et al.*, 1997), which assumes that the closer the flow regime is to the natural regime, the better it is for the environment. The approach described in this paper does not rely on that assumption. Rather, the preferred environmental flows were specified as a regime that would maintain the ecological assets identified in the river at a given level of risk. In evaluating the trade-off between reliability of supply and meeting ecological needs, modelled system operation scenarios were used to assess the degree to which the individual components of a preferred (low risk) and a sub-optimal (higher risk) environmental flow regime were met, rather than measuring their proximity to the natural regime.

The Taizhou study described here was undertaken to demonstrate the process of water resources allocation planning, and the likely consequence of adopting alternative management arrangements. The main objectives were to iteratively generate a range of potential water resources allocation plan scenarios, and use these scenarios to evaluate and demonstrate the trade-offs between meeting security of supply and environmental flows objectives. This investigation was concerned with comparison of the relative performance of the scenarios rather than optimization of detailed reservoir operating rules for each scenario. This paper describes a risk assessment approach that can be used to derive sub-optimal environmental flow regimes from a preferred regime. Also, a new approach is presented for assessing a flow series for its degree of compliance with a specified environmental flows regime.

Background to the Taizhou Case Study

Taizhou is a prefecture-level city in eastern Zhejiang Province (Figure 1) with a population of 5.47 million people. Zhejiang Province is undergoing a period of rapid development which, combined with increasing standards of living, will lead to increased demand on water supplies. A comprehensive water resources plan for Taizhou Prefecture was finalized in 2004 (Taizhou Water Resources Bureau, 2004). This was prepared in accordance with central- and provincial-level requirements and in the context of water shortages that occurred throughout Taizhou in 2003.

Total water demand in the south of Taizhou (which is currently largely supplied by the Changtan Reservoir) (Figure 2) is expected to increase from $76.5 \times 10^6 \, m^3$ in 2010 to $117.7 \times 10^6 \, m^3$ in 2020. Most of the additional demand will come from growth in urban and industrial consumption. The increase in urban consumption will result from population growth as well as predicted increases in average domestic use from improved living standards. Currently there are four major storages in Taizhou (Figure 2). The largest is Changtan Reservoir, built in the period from the late 1950s to the early 1960s on Yongningjiang. This storage has an effective storage volume of $356 \times 10^6 \, m^3$, mean annual total inflow of $564 \times 10^6 \, m^3$, and is used predominantly for urban supply to Jiaojiang District. The other main storages are Niutoushan Reservoir (built 1989, capacity $302 \times 10^6 \, m^3$), Lishimen Reservoir (capacity $199 \times 10^6 \, m^3$) and Xia'an Reservoir ($135 \times 10^6 \, m^3$). Two major new storages, the Zhuxi Reservoir (proposed capacity $125 \times 10^6 \, m^3$) and Shisandu Reservoir (proposed capacity $145 \times 10^6 \, m^3$), have been proposed for tributaries of Yong'anxi in the southwest of Taizhou to accommodate projected increases in water demand (Figure 2). The two new storages, when combined with the Changtan Reservoir, are expected to provide sufficient supplies for development in the south of Taizhou beyond 2020.

Zhuxi Reservoir is due to be operational by 2011. It will be located on Zhuxi (Zhu Creek) in Xianju County (Figure 2). The construction of two or three new weirs on the tributary downstream of the dam is also proposed. The catchment area at the dam site is $172.3 \, km^2$ and the mean annual flow is $192 \times 10^6 \, m^3$. The area to be inundated by the reservoir will be about 460 ha. The dam wall will be about 70 m high.

Figure 1. Location of Taizhou study area in eastern Zhejiang Province, People's Republic of China.

The effective storage volume will be $93.5 \times 10^6 \, m^3$. The vast majority of water captured by the dam will be piped to Changtan Reservoir, for supply to the southern regions of Taizhou, where water shortages are critical. The dam will also serve a flood mitigation function, and will be used for opportunistic hydropower generation. The construction of Shisandu Reservoir is subject to future assessments of Taizhou's water demand.

There is a general expectation among stakeholders that urban and industrial users will receive a daily reliability of supply of 95 or 97%, while agricultural users will receive annual reliability of supply of 75%. Designs for future water supply projects in the People's Republic of China typically aim to achieve these reliabilities. In general, a demand-driven approach has been taken to water planning in Taizhou, with minimal consideration of supply limitations. Planning has tended to focus on meeting existing and future demands through engineering solutions, rather than improving the allocation of existing supplies through efficiency measures, reallocation of surplus between sectors, or water trading. There is currently no clearly defined cap controlling the number of permits that can be granted or the total volume of water that can be abstracted from individual

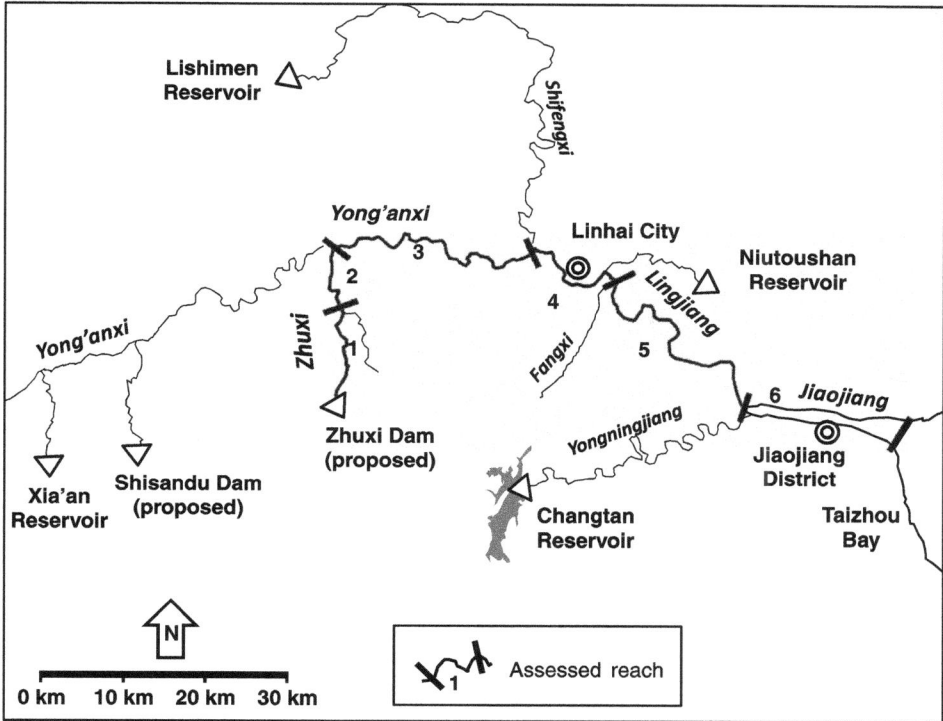

Figure 2. The six study reaches of the Jiaojiang Basin, Taizhou, considered in the environmental flows assessment. *Note:* Not all streams and reservoirs are depicted.

regions or the basin as a whole, which places the reliability of supply at risk (Water Entitlements and Trading Project (WET), 2007).

It would appear that considerable discretion exists for how available water is shared amongst users when in short supply. Should such circumstances arise, specific arrangements are developed between the county and prefecture governments. In the absence of water sharing arrangements, entitlement holders are allowed to take as much water as they want (up to their entitlement limit) on a first-come first-served basis. As dam levels approach contingency cut-off points, this can encourage individuals to use their water as fast as possible, to avoid missing out, and thus speed the region towards a contingency situation. Contingency measures are only applied when water supplies are at a critical level. There are no efforts to reduce consumption (e.g. low-level restrictions) by any sector until then. This means that when restrictions are eventually imposed, they can be severe and abruptly implemented.

While prioritization of various users is apparent, it is not clear what (if any) provision has been made for natural ecosystem requirements in the assessment of storage yields and reliabilities. Currently, it appears that consideration is given to releasing water from storages, or allowing flows to reach the end of river systems, on an ad hoc basis only. Further, any such decisions appear to be made in order to dilute downstream pollutants and for off-stream consumption associated with beautification, restoration, dust suppression or other landscape amelioration activities, rather than to meet the ecological needs of the rivers. One reason for the lack of consideration of in-stream ecological needs is that they have not previously been defined. A recent environmental flows assessment has provided the necessary information for four freshwater reaches of the Jiaojiang Basin (Figure 2) (Gippel *et al.*, 2009). The two defined estuarine reaches of the river (Figure 2) were investigated by Gippel *et al.* (2009), but reach-specific environmental flows recommendations were not made.

Methodology

Water resources allocation planning aims to set abstraction caps and protect the reliability of supply for water users, while also providing for environmental flows needs. The study described here focussed on the proposed Zhuxi Reservoir, for which environmental flows recommendations were available (Gippel *et al.*, 2009). The reliability of supply from Changtan Reservoir was also modelled because this would likely be affected by diversion of water from Zhuxi Reservoir to Changtan. Thus, the study considered the impact of different management arrangements on, and what different options for a water resources allocation plan would mean for:

- reliability of local supply from the proposed Zhuxi Reservoir;
- reliability of supply from Changtan Reservoir; and
- impacts on river flows, from the Zhuxi dam site downstream to the estuary.

The water resources allocation planning methodology adopted in this study involved the following seven steps:

1. Identification of water resources availability and supply requirements.
2. Identification of environmental flows requirements.
3. Development of a water resources management model (IQQM: Integrated Quality and Quantity Model), capable of modelling the impacts of different operational and supply arrangements on the river flow and supply reliability.
4. Modelling of the current management and supply arrangements.
5. Development of different management scenarios, including scenarios that incorporate environmental flows requirements.
6. Analysis of the modelled results for each scenario, in terms of the impact on flows deemed important for the environment, and reliability of supply to water users (assuming fixed demand).
7. Identification of the way that a preferred management scenario could be converted into a water resources allocation plan.

The final step 7 (above) is not discussed in this paper, but the details of that step as applied to the Taizhou study are provided in WET (2007). This paper does not make a final recommendation on a particular preferred planning scenario. The decision on what are acceptable outcomes, in terms of water supply reliability and risks to the environment, is a matter for the local government to decide.

Water Resources Allocation Plan Scenarios

Various scenarios were developed to allow assessment of the impacts of different management arrangements on the provision of water for environmental purposes and human uses. Basic scenarios were developed to compare the current situation with the arrangements proposed for the year 2020 (including the new Zhuxi and Shisandu Reservoirs), and also for the proposed 2020 arrangements with environmental flows rules incorporated. Further scenarios were developed to assess other issues, including the effects of different operational arrangements for the proposed infrastructure, alternative develop-ment proposals, and various combinations of these scenarios. The process of developing the scenarios was iterative, allowing for the consideration of the advantages and disadvantages of each scenario with a view to providing viable options for a water resources allocation plan. Eleven different scenarios were modelled (Table 1, Figure 3).

The scenarios were largely based on the assumed demands for the year 2020 that were outlined in the Comprehensive Plan for Water Resources in the Taizhou Prefecture (Taizhou Water Resources Bureau, 2004) (except for Scenario S1, which was the current situation base case, and Scenario S7). Scenarios that utilized environmental flows rules adopted the preferred flow regime as specified in Gippel *et al.* (2009). The exception was Scenario S11, which modelled the impact of a sub-optimal environmental flows regime developed by the expert panel using a risk assessment procedure (described below). This scenario was designed to consider the implications of a 'trade-off' of some environmental outcomes in favor of improved supply reliabilities. The modifications to the operating arrangements for different infrastructure (Scenarios S4, S5, S9, S10 and S11) were considered as possible ways of improving supply reliabilities for various users—both with and without the provision of environmental flows. Scenarios S6 and S10, which contemplated an additional hypothetical reservoir being constructed in the Yong'anxi catchment upstream of Zhuxi (Figure 4), were included to demonstrate the cumulative impacts of continuous development. Scenario S7 assessed the capacity to meet demands for the year 2010 from existing infrastructure (i.e. without Zhuxi Reservoir). Scenario S8 assessed whether the expected demands for the year 2020 could be met without transferring water from Shisandu Reservoir to Zhuxi.

Table 1. Modelled water resources allocation plan scenarios

Scenario	Description
S1	Existing (2004) conditions
S2	Planned 2020 arrangements (demand based on Taizhou Comprehensive Plan; assumes construction of Zhuxi and Shisandu Reservoirs)
S3	Same as S2, but with the preferred environmental flows rules as recommended from the environmental flows assessment
S4	Same as S2, but with modified Changtan Reservoir operation rules (altered sharing rule for urban and irrigation supply)
S5	Same as S4, but with preferred environmental flows rules as recommended from the environmental flows assessment
S6	Same as S2, but with additional reservoir with local demands equivalent to Zhuxi Reservoir in the Yong'anxi catchment upstream of Zhuxi
S7	2010 demands with no Zhuxi Reservoir and no environmental flows rules
S8	Same as S2, but with no water transfers from Shisandu Reservoir
S9	Same as S2, but with lower release threshold for Zhuxi Reservoir water transfers to Changtan Reservoir
S10	Same as S5, but with additional reservoir with local demands equivalent to Zhuxi Reservoir in the Yong'anxi catchment upstream of Zhuxi
S11	Same as S4, but with modified (sub-optimal) environmental flows rules

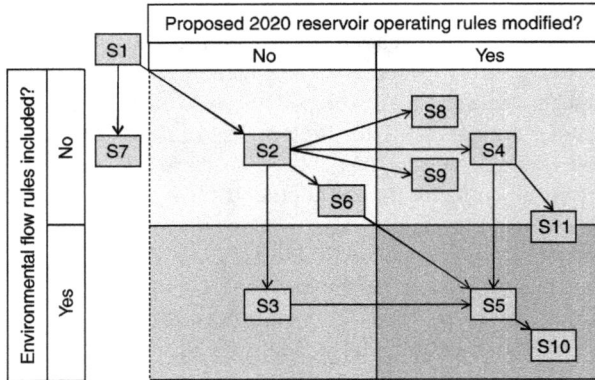

Figure 3. Diagrammatic representation of modelled water resources allocation plan scenarios. *Note:* Scenario S1 represents the existing (2004) conditions; S7 is for year 2010 demands and others are for year 2020; S11 uses a sub-optimal environmental flows regime; other scenarios are variants of one of the four indicated categories.

Water Resources Management Model

Water resources modelling was done using IQQM, which is a hydrologic modelling tool developed in Australia for use in planning and evaluating water resource management policies (Department of Land and Water Conservation, 1998). The IQQM was used to model each of the scenarios over the 27-year simulation period from 1980 to 2006, with results reported at six nodes (Figure 4). Details of the model parameters and assumptions are provided in WET (2007). The statistics of interest for each scenario were:

- Daily and annual reliabilities (for fixed supply volumes) for urban/industrial users and for irrigation, from both Zhuxi and Changtan Reservoirs.
- Average annual supply for urban/industrial users and irrigation, both as a volume and a proportion of the sectors' demand, from both Zhuxi and Changtan Reservoirs.
- Average annual transfer of water from Zhuxi Reservoir to Changtan Reservoir, both as a volume and a proportion of the maximum possible through the pipeline.

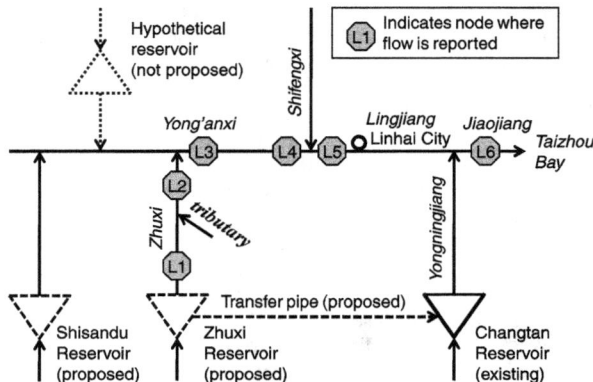

Figure 4. Schematic of main IQQM model components, and reporting nodes

- Time series of mean daily discharge for the reporting nodes, which was used to assess compliance with the preferred environmental flows regimes.

The recommended environmental flows requirements (Table 2) were incorporated as rules into a number of model runs (Table 1, Figure 3). While it would have been technically feasible to incorporate into the IQQM model all of the flow recommendations for all four assessed reaches, this was not done because of the project's focus on Zhuxi Reservoir. In reality, all of Taizhou's reservoirs and rivers will have to be managed in an integrated way, but the model did not attempt to fully simulate this process. The focus on Zhuxi Reservoir meant that the flow rules (Table 2) for the two Zhuxi reaches (Figure 2) were incorporated into the IQQM model, as these were under the direct control of the reservoir. The flows in Yong'anxi (Figure 2), although not intentionally manipulated in the IQQM model to meet environmental flows requirements, would gain an incidental benefit from the environmental flows released to Zhuxi.

Generation of Sub-optimal Environmental Flows Regime Options

Environmental flows normally reduce security of supply, so it is often the case that managers seek a recommendation from scientists on alternative (sub-optimal) environmental flows regime options that will achieve higher security of supply. Reducing the environmental allocation requires prioritization of the identified assets and objectives to determine where reductions in the environmental allocation are best made. One expedient approach would be to remove entirely from the recommended regime one or more of the flow components that require large volumes of water to implement (typically, the larger pulses and the channel maintenance events would be targeted). However, this means that the assets that relied on those components would be placed under serious risk. An alternative strategy, explored here, is to modify the facets of the recommended flow components—the magnitude, frequency and/or duration—in ways that improve reliability of supply for consumptive use, but minimize the risk to the environment. It must be clearly recognized that any sub-optimal environmental flows option represents a departure from the recommended (preferred) flow regime (Table 2), so it would carry a higher risk that the identified assets will not be protected. Accepting a higher risk is a management decision, not a scientific one.

In providing an alternative set of environmental flows recommendations, there are a number of options to consider. The relative potential to modify (i.e. reduce) a flow component depends on two things: (i) the relative potential to improve security of supply with that reduction, and (ii) the relative risk to the environment of undertaking that reduction. If reducing a flow component has negligible impact on security of supply then there is little point in taking the environmental risk associated with the reduction. However, if reducing a flow component carries a high risk to the environment, then the reduction should not be considered.

The relative potential to improve security of supply was determined through an initial modelling exercise by calculating the volumes of water required to supply the various flow components (i.e. volume = frequency × magnitude × duration), and then determining the relative impact on total system water requirements by reducing the volume of each flow component by half. For example, High flows are of a relatively low magnitude but long duration, so they account for a large percentage of the total flow volume required; Overbank flows are infrequent but of high magnitude, so they also account for a high

Table 2. Specifications of the preferred environmental flows components

Component	Timing	Magnitude ($m^3 s^{-1}$)	Duration	Annual frequency	Long-term period frequency
Reach 1: Zhuxi immediately downstream of dam site					
LF	Oct–Mar	0.5	65% of time	–	5 in 10 years
HF	Apr–Sep	1.3	65% of time	–	5 in 10 years
LFP	Oct–Mar	20	1 day	2	5 in 10 years
HFP1	April	20	1 day	1	5 in 10 years
HFP2	May–Sep	20	2 day	4	5 in 10 years
BF	Jan–Dec	177	1 day	0.5 ARI	3 in 10 years
OB	Jan–Dec	191	1 day	0.5 ARI	3 in 10 years
Reach 2: Zhuxi upstream of Yong'anxi junction					
LF1	Oct–Jan	1.3	65% of time	–	5 in 10 years
LF2	Feb–Mar	3.4	65% of time	–	5 in 10 years
HF	Apr–Sep	4.0	65% of time	–	5 in 10 years
LFP	Oct–Mar	53	1 day	2	5 in 10 years
HFP1	April	53	1 day	1	5 in 10 years
HFP2	May–Sep	53	2 days	3	5 in 10 years
BF	Jan–Dec	397	1 day	0.5 ARI	3 in 10 years
OB	Jan–Dec	428	1 day	0.5 ARI	3 in 10 years
Reach 3: Yong'anxi upstream of Shifengxi junction					
LF1	Oct–Jan	6.8	65% of time	–	5 in 10 years
LF2	Feb–Mar	17	65% of time	–	5 in 10 years
HF	Apr–Sep	22	65% of time	–	5 in 10 years
LFP	Oct–Mar	96	2 days	3	5 in 10 years
HFP	April–Sep	146	4 days	3	5 in 10 years
BF	Jan–Dec	948	1 day	0.5 ARI	3 in 10 years
Reach 4: Yong'anxi upstream of Fangxi junction					
LF1	Oct–Jan	16	65% of time	–	5 in 10 years
LF2	Feb–Mar	32	65% of time	–	5 in 10 years
HF	Apr–Sep	46	65% of time	–	5 in 10 years
BF	Jan–Dec	2836	1 day	0.5 ARI	3 in 10 years

Note: Only events meeting all the criteria comply with the specifications.
LF = Low flow; HF = High flow; LFP = Low flow pulse; HFP = High flow pulse; BF = Bankfull; OB = Overbank; ARI = Average recurrence interval.

percentage of the total flow volume required. Based on the calculated volume of water required, each of the components was scored for potential to improve security of supply if it was reduced. A scale over the range 0–3 was used (0 = Nil, 1 = Low, 2 = Moderate, 3 = High, with intermediate scores possible).

The relative environmental impacts of reducing the facets (i.e. magnitude, duration and frequency) of each of the flow components were determined by a simple risk assessment. The risk was based on a number of principles: (i) Reducing a component's magnitude carried the greatest risk (as the necessary hydraulic threshold might not be achieved); (ii) Low flow and High flow magnitudes are low in the channel, and a significant reduction would likely create a major loss in available habitat. However, the high flow season was considered less important than the low flow season in this respect; (iii) Bankfull and Overbank are less frequent than Pulses, so reducing their frequency was considered more detrimental than reducing the frequency of Pulses (which would still occur at least once per year); (iv) Pulse duration was considered marginally more important than frequency, as a certain duration is needed in order to allow the ecological process to be completed (i.e. water could be wasted if an inadequately short event was released, so it would be preferable to save the water for fewer, but effective, events); and (v) Bankfull and Overbank components were specified with a peak duration of only one day, so total event duration could not be reduced without increasing the rate of rise and fall. On the basis of these principles, each facet of each component was assigned a score for relative risk to environment (High = 0, Moderate = 1, Low = 2, Nil = 3, with intermediate scores also possible).

To achieve a ranking of potential to modify (i.e. reduce) the facets of the flow components, a risk assessment was undertaken by multiplying the score for potential to improve security of supply by the score for relative risk to environment, and then ranking

Table 3. Relative potential to modify the facets of the environmental flows components

Facets of the flow components	Relative security of supply improvement	Relative risk to environment	Relative potential to modify	Rank potential to modify
LF magnitude	Moderate	High	Nil	–
HF magnitude	Mod–High	Moderate	Moderate	4
LFP magnitude	Low	High	Nil	–
LFP duration	Low	Moderate	Low	7
LFP frequency	Low	Low	Low–Mod	5
HFP magnitude	Moderate	High	Nil	–
HFP duration	Moderate	Moderate	Low–Mod	6
HFP frequency	Moderate	Low	High	1
BF magnitude	Mod-High	High	Nil	–
BF duration	Nil[a]	Moderate	Nil[a]	6[a]
BF frequency	Mod–High	Moderate	Moderate	3
OB magnitude	High	High	Nil	–
OB duration	Nil	Moderate	Nil	–
OB frequency	High	Moderate	Mod–High	2

Note: a. Reach 3 an exception, with moderate potential to improve security of supply. Implement with High flow pulse duration reduction. LF = Low flow; HF = High flow; LFP = Low flow pulse; HFP = High flow pulse; BF = Bankfull; OB = Overbank.

the facets of the components according to the total risk score (Table 3). A number of rules were followed in deriving the final rankings:

1. The lower the risk to the environment and the higher the potential for improving security of supply, the higher the 'relative potential to modify'.
2. Reduction in event magnitude resulted in High risk to the environment due the failure to achieve the threshold required to initiate the ecological or geomorphological processes.
3. A High risk to the environment resulted in no potential to modify the component.
4. A Nil potential to improve security of supply resulted in no potential to modify the component.
5. Reduction in Pulse frequency was favoured over reduction in duration.
6. Pulse durations of one day could not be reduced.
7. Bankfull event duration could only be reduced for Reach 3, and Overbank duration could not be reduced.
8. The procedure for ranking options to modify the hydrological facet of flow components did not allow sequential ranking within a component, i.e. if reduction in frequency of a component ranked highly, the duration of that component could not be ranked immediately after it. The intention of this was to reduce the likelihood that the final alternative environmental flows regime would involve a reduction of both duration and frequency to any one component.
9. Facets with equal risk scores were given a rank order according to their relative risk to environment score.

The final rank of potential to modify (Table 3) enabled the generation of alternative environmental flows options. As additional changes were progressively included (in the order indicated by the rank) the potential to achieve an improvement in security of supply increased but so too did risk to the environment. These potentials and risks are not numerically related and must be evaluated qualitatively. For this exercise one alternative sub-optimal environmental flows option was generated for inclusion in Scenario S11 (Table 1, Figure 3). This option included all seven ranked potential changes, so it can be regarded as the highest risk to the ecological assets that the scientific panel was prepared to take, short of eliminating flow components in their entirety. For each of the seven facets that were changed, the change was to halve the value of the facet. The exception was High flow magnitude, which the expert panel felt carried excessive risk to the environment by halving, so it was reduced by 25%. The High flow pulse associated with spawning was regarded as high risk to the environment if reduced in any way, so it was not modified. For this exercise, the Overbank component was removed from consideration. Overbank would currently achieve little more than Bankfull, as most of the potential floodplain assets in the Jiaojiang system are protected by levees. Overbank would be a realistic option only in the event that some areas of floodplain were to be re-connected.

Evaluating Compliance with Environmental Flows Requirements

Compliance is the degree to which the specified flow components occur in the flow series. To comply with the requirements of a component, an event-type component (i.e. Pulse, Bankfull or Overbank) has to satisfy the specifications for all three facets—frequency, duration and magnitude. The compliance for events was calculated using a sophisticated form of spells analysis. In calculating the occurrence of events, independence was defined by a period between events of seven days, provided the flow did not fall by more than 25% from the threshold, and all calculations assumed a water year beginning in January. The event was deemed to have occurred only when the magnitude was exceeded for the required duration. The frequency of occurrence of the events was then summed for each year of record. Each year was then assessed for the minimum number of times the component was required to occur. If there were too few events then that year was non-compliant. Having more than the minimum recommended number of events in a particular year did not result in 'over-compliance' that could somehow compensate for other years when the event did not occur frequently enough. One potential measure of compliance is the percentage of years in the record that satisfied all of the annual requirements of the component. The closer that value was to 100%, the higher the level of compliance. A value of zero would mean that the component did not occur in the time series. Not all components were expected to have a compliance of 100% with the preferred environmental flow regime in the current scenario. There are three main reasons for this:

1. Most river reaches are already regulated to some degree (the exception is Zhuxi at the dam site).
2. The time series contained drought years when the flow components would not necessarily be expected to occur.
3. The hydraulic models contained uncertainties that could have led to inaccuracies in the specification of flow magnitudes required to achieve the given ecological and geomorphological objectives.

Given these factors, for event-type components, the flow series could be said to 'comply' if the event was satisfied in more than a certain percentage of years (by setting an arbitrary threshold). A weakness of this approach is that it takes no account of the temporal distribution of the non-complying years. For example, a long sequence of non-complying years for a fish spawning component could be catastrophic for the species in question. Thus, a long-term frequency requirement was specified for each component (Table 2). This was stated as the number of years in every 10 years that the component had to comply. In this sense, 'every 10 years' means every sequence of rolling 10-year long periods in the record, not simply the record divided into discrete periods, each of 10 years length (Figure 5). The period length does not have to be 10 years—the environmental flows scientific panel uses their collective expertise to establish a meaningful period length and required frequency. For the Jiaojiang system, a frequency of five complying years in every rolling 10-year period was specified for most components. For the less frequent Bankfull and Overbank components the requirement was three complying years in every 10 (Table 2). For example, if a pulse component appeared in at least five years in every one of the 18 sequences of 10 years in the 27-year modelled period, then the compliance was 100%, while compliance of 50% would mean that in nine of the 18 rolling 10-year periods the component appeared in at least five of those 10 years (Figure 5).

Compliance for Low flow and High flow (baseflow) components had to be calculated using a slightly different approach, because baseflows are not specified with a frequency. For baseflows, duration above the threshold is the important characteristic. A required duration of 65% of the time over the relevant season was set as the lower limit of annual compliance for each baseflow component (this percentage was based on expert opinion, and would likely vary between river systems) (Table 2). Due to the high degree of variability of flow from year

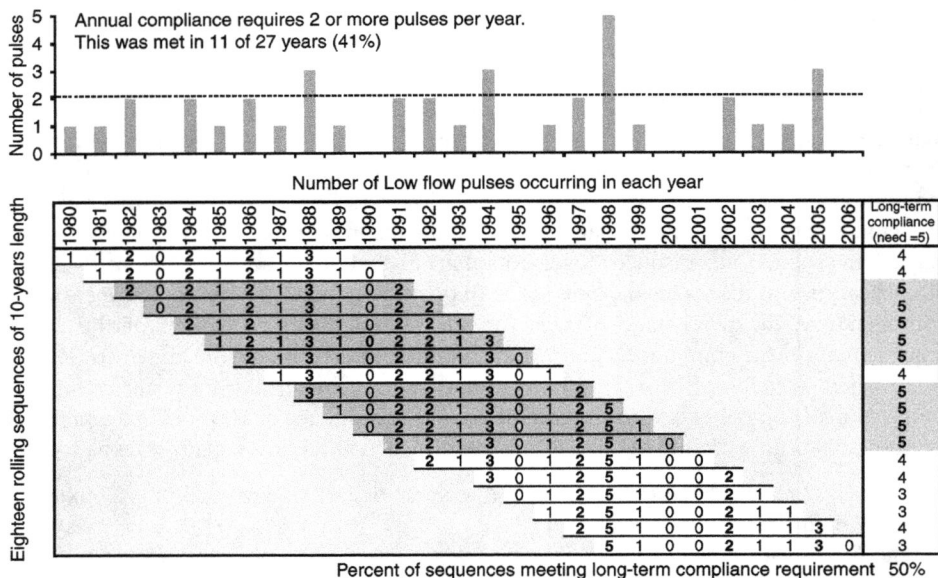

Figure 5. Illustration of calculation of long-term compliance of 50% for the component Low flow pulse at Reach 2: Zhuxi upstream of Yong'anxi junction, under Scenario S3: with dams and preferred environmental flows releases

to year, it was only expected that the threshold baseflow duration requirement would be met in five years of every rolling 10-year period (i.e. the other years were too dry to meet this requirement). In rivers with low inter-annual hydrological variability, a higher expectation can be placed on the long-term frequency of flow components.

Results

Degree of Compliance with Environmental Flows Recommendations

The degree of compliance with the preferred environmental flows recommendations was assessed for three scenarios, Scenario S1 (current), Scenario S2 (Year 2020 with Zhuxi and Shisandu Reservoirs operational), and Scenario S3 (as for scenario S2 but with environmental flows releases) (Figure 6). These three scenarios captured the majority of the potential range in compliance across all of the scenarios. For example, Scenario S2 had the same level of compliance as Scenarios S4 and S6, and similarly Scenario S3 produced the same compliance result as Scenario S5.

The current Scenario S1 had the overall highest level of compliance (Figure 6). This was expected, as the current series was utilized to help specify the frequency and duration of the recommended flow components. Scenario S2, which had the dams in place but no allowance for environmental flows, had very low compliance, particularly for the lower magnitude flow components. The compliance of Scenario S2 improved with distance from Zhuxi dam, but even then, the High flow component was poorly represented. Mean annual flows as a percentage of current flows were 42% at Zhuxi DS (downstream) of dam site, 73% at Zhuxi US (upstream) of Yong'anxi, 82% at Yong'anxi US of Shifenxi and 89% at

Reach 1: Zhuxi downstream of Dam site

	S1	S2	S3
■LF	100%	0%	100%
■HF	100%	0%	100%
■LFP	100%	0%	100%
■HFP1	78%	0%	100%
■HFP2	100%	0%	100%
BF	100%	100%	100%
OB	89%	100%	100%

Reach 2: Zhuxi upstream of Yong'anxi

	S1	S2	S3
■LF1	100%	0%	100%
■LF2	100%	33%	100%
■HF	94%	0%	83%
■LFP	72%	0%	50%
■HFP1	94%	50%	83%
■HFP2	100%	67%	100%
BF	78%	100%	83%
OB	78%	100%	78%

Reach 3: Yong'anxi upstream of Shifengxi

	S1	S2	S3
■LF1	100%	11%	44%
■LF2	100%	89%	100%
■HF	89%	0%	39%
■LFP	100%	89%	100%
■HFP	100%	78%	83%
BF	100%	100%	100%

Reach 4: Yong'anxi upstream of Fangxi

	S1	S2	S3
■LF1	100%	78%	94%
■LF2	100%	100%	100%
■HF	100%	28%	78%
BF	83%	83%	83%

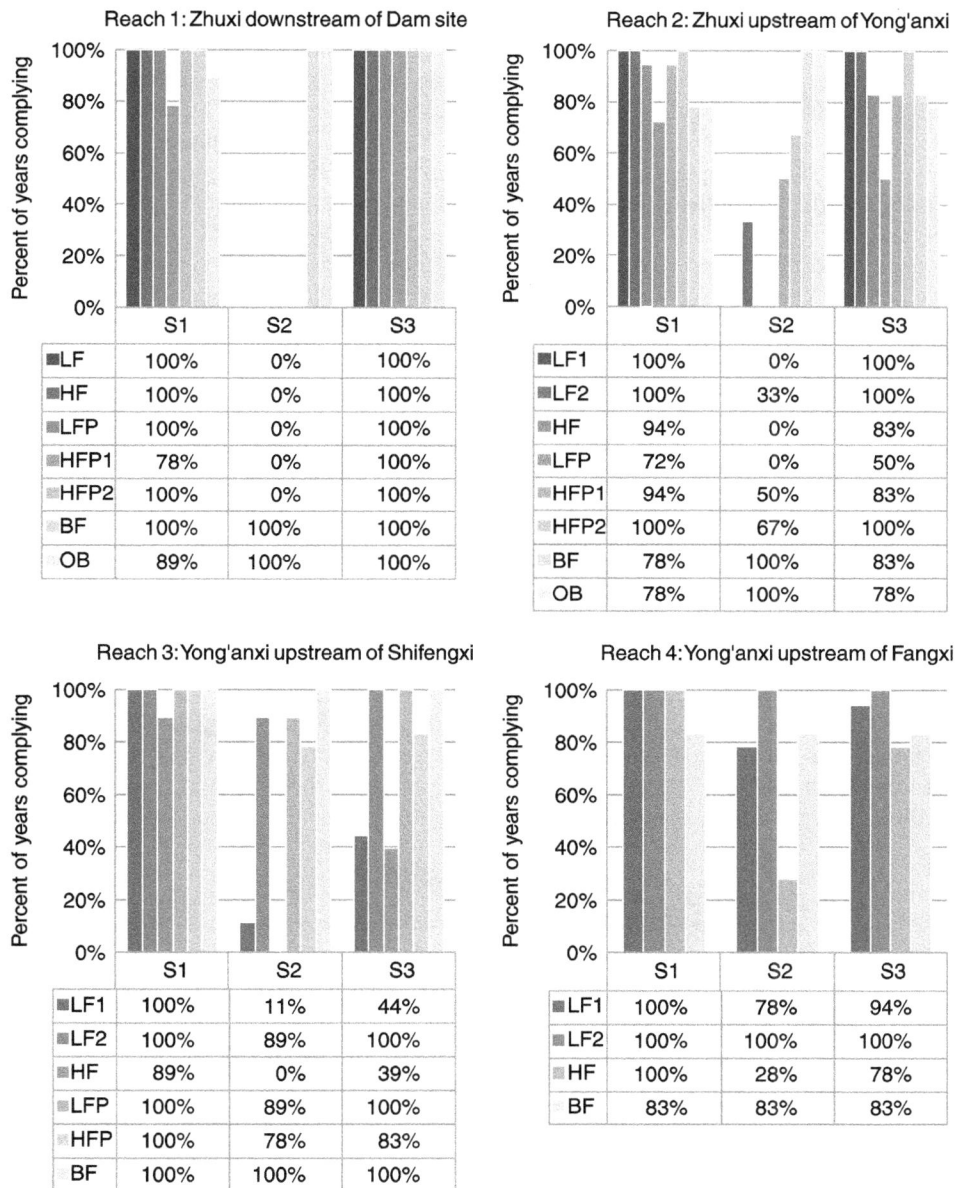

Figure 6. Compliance of modelled flow scenarios with preferred environmental flows requirements. *Note:* S1 = Existing conditions (year 2004); S2 = Planned year 2002 arrangements with Zhuxi and Shisandu Reservoirs operational; S3 = Same as S2 but with preferred environmental flows rules implemented. Compliance of components was measured over a 27-year modelled period, and took account of magnitude, annual frequency, annual duration and long-term frequency.

Yong'anxi US of Fangxi. Incorporating environmental flows in Scenario S3 dramatically improved compliance (Figure 6). Downstream of the dam the compliance was 100% for each component, which was expected, as the dam was specifically managed to achieve this result. Mean annual flows as a percentage of current flows were 67% at Zhuxi DS of Dam site, 83% at Zhuxi US of Yong'anxi, 84% at Yong'anxi US of Shifenxi and 90% at Yong'anxi US of Fangxi. Operation of the dam to deliver environmental flows had a significant effect on drawing the post-dam flow duration curve closer to the pre-dam curve (Figure 7).

Zhuxi at its lower end achieved good compliance under Scenario S3, except for low flow pulses. Yong'anxi upstream of Shifengxi had good compliance except for early dry season Low flows and High flows. The lower reach of Yong'anxi had acceptable levels of compliance (Figure 6). Bankfull and Overbank flow components had acceptable compliance in all scenarios at all compliance points (Figure 6). These components occur mainly in response to typhoon rains that cause uncontrolled dam spills, so to a large extent they are beyond management control.

The degree of compliance of Scenarios S1, S2 and S3 was also measured against the requirements of the sub-optimal environmental flow regime. As expected, the degree of compliance with this regime was higher, but these criteria were associated with a high risk that the health of the ecological assets would not be maintained. Nevertheless, under these criteria, for Scenario S3, all four reaches had greater than 90% compliance for all flow components except for Low flow 1 at Reach 3 (which remained at 44% compliance). Scenario S2 had poor compliance with the sub-optimal flows regime for both reaches of Zhuxi, clearly indicating the high risk associated with this scenario.

Overall, the model results suggested that incorporating the recommended environmental flows rules into dam operation would provide a reasonably high level of protection for the ecological health of Zhuxi, and significant incidental benefits for Yong'anxi. This analysis also provided a dramatic illustration of the potential impacts on the health of the

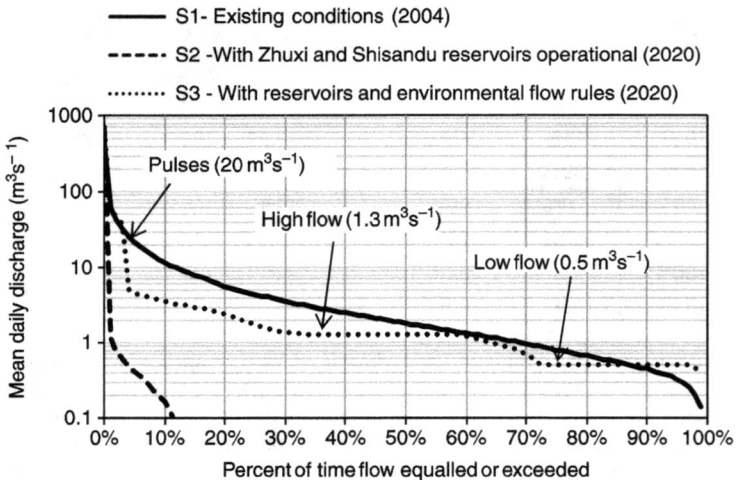

Figure 7. Daily flow duration curves for three modelled scenarios for the site immediately downstream of the proposed Zhuxi Dam. *Note:* High flow and Low flow components have a strong influence on the shape of the curve.

river system if environmental flows releases were not to be incorporated into the operation of Zhuxi Reservoir. The consequences for Zhuxi, and to a lesser extent Yong'anxi, would likely include:

- loss of habitat (especially the loss of deep pools);
- loss of pulses to trigger spawning events;
- loss of connectivity for migration;
- loss of flows to maintain riparian vegetation; and
- incidents of poor water quality.

Together, these changes would likely have a major impact on fish and fisheries within Zhuxi and further downstream. From Yong'anxi downstream to the estuary, the flow regime would likely deteriorate (from the perspective of its suitability for supporting a healthy ecosystem) over time in response to the cumulative impacts of basin-wide water resources development.

Reliability of Supply

Under the proposed arrangements for year 2020 (Scenario S2), local water supply from Zhuxi Reservoir for urban/industrial uses would be provided at daily reliability of 97%, while annual reliability for irrigation would be 100% (Figure 8). From Changtan Reservoir, water supply for urban/industrial users would be provided at a daily reliability of 82%, while annual reliability for irrigation would be 0% of years with full demand met (Figure 9). Although annual reliability for full supply would be poor, on average, 69% of the annual demand volume would be met. Notably, this shows that under the currently proposed 2020 arrangements reliability of supply for all purposes from Changtan Reservoir is likely to be below the acceptable levels.

The results obtained for Scenario S2 prompted the consideration of a number of scenarios (S4, S5, S10 and S11) involving a lower threshold (i.e. defined water level in Changtan Reservoir) for restricting urban/industrial water supplies (the threshold for irrigation supplies was not changed), essentially allowing the reservoir to be drawn down further than would currently be permitted. These modified operation rules would significantly improve the daily reliability of supply for urban/industrial users to a level around the desired 95% (depending on the other factors in the scenarios), with slight consequences for the average annual supply for irrigation users from Changtan Reservoir (Figure 9). A second possible method for pursuing higher supply reliabilities, through amended infrastructure operating rules, was also considered (Scenario S9), but this had only minor positive impacts (Figure 9) and significant negative consequences for irrigation supply reliability from Zhuxi Reservoir (Figure 8).

Provision of environmental flows (Scenarios S3, S5 and S10) had mixed impacts on reliability of supply (Figure 8 and Figure 9). Urban/industrial users from Zhuxi Reservoir would have no reduction in reliability of supply compared to the comparable scenarios without environmental flows. However, irrigation users would see average annual supply fall from 100% to 95% of their demand, and annual reliability of supply fall from 100% to 67% (Figure 8). Provision of environmental flows would also reduce the reliability of urban/industrial supplies from Changtan Reservoir (Figure 9). This comes about because opportunities to transfer water from Zhuxi Reservoir to Changtan Reservoirs via the proposed connecting pipe (Figure 4) depend on having high water levels in Zhuxi

Reservoir, and environmental flows releases tend to maintain lower water levels in the storage. With environmental flows releases from Zhuxi Reservoir, the average annual supply from Changtan would be reduced by 1–3% of the urban/industrial sector's demands (depending on other factors considered in the scenarios). A comparison of Scenarios S2 and S3 indicated that daily reliability would be significantly reduced (from 82% to 74%), unless the Changtan Reservoir operation rules were also changed in the manner described above (Scenarios S4 to S5), in which case the reduction would be relatively minor (from 96% to 93%). The annual average supply for irrigation users from Changtan Reservoir would be between five and seven percentage points lower with environmental flows rules in place (to between 60% and 64% of demand, with no years where full demand would be met).

Cumulative Impacts of Development

Two scenarios (S6 and S10) were modelled to demonstrate the cumulative effects of development on flows downstream. These scenarios both contemplated an additional (hypothetical) reservoir being constructed in the Yong'anxi catchment upstream of the point where Zhuxi meets Yong'anxi (Figure 4). The modelling assumed identical levels of local demand as from Zhuxi Reservoir but, significantly, did not involve the transfer of water out of the catchment (i.e. it was assumed this reservoir would only supply relatively small local

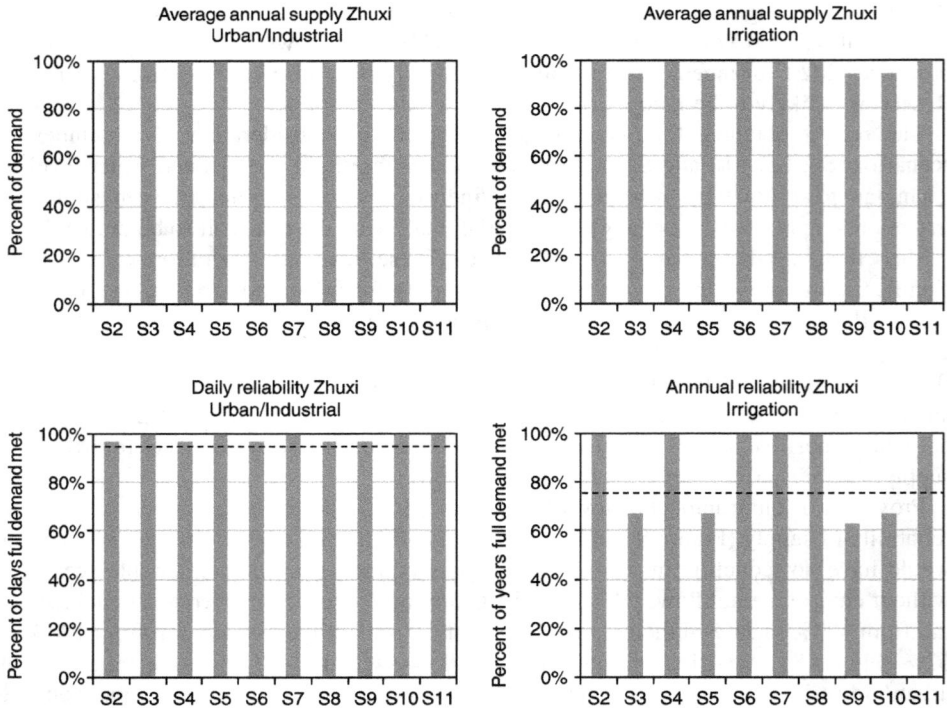

Figure 8. Modelled supply reliabilities for Zhuxi Reservoir urban/industrial and irrigation users. *Note:* Dashed lines show expected levels of reliability of supply.

demands). Scenario 6 was based on the proposed 2020 arrangements. As such, the results for this scenario should be compared to those for the current planned development (Scenario S2). Scenario S10 was based on the 2020 arrangements with modified Changtan Reservoir operating rules and the provision of environmental flows. The model results for this scenario should be compared to the results from Scenario S5.

As the new hypothetical reservoir was not connected to the Zhuxi and Yongningjiang systems, there were no implications for supply from Zhuxi or Changtan Reservoirs, nor for flows (and thus the achievement of environmental outcomes) within Zhuxi. As such, the comparison of modelling results only needed to focus on flows in Yong'anxi, Lingjiang and Jiaojiang, i.e. at IQQM reporting locations L3 to L6 (Figure 4). The impact of this hypothetical development was relatively small, reducing the proportion of existing mean annual flow in the river at the upstream node (L3) by 0.8% of the total flow and 0.3 to 0.4% of the total flow at the downstream node (L6). However, this impact would have been greater if the additional reservoir was used to supplement water supplies in other parts of Taizhou as well as meeting local demands.

Exploring Less Intense Development Options

Two additional scenarios (S7 and S8) were modelled to consider the implications of not building some of the infrastructure planned for in the Taizhou Comprehensive Plan. The first (Scenario S7) assumed Zhuxi Reservoir was not in fact built by 2010. The model results indicated that the predicted 2010 demands could generally be met

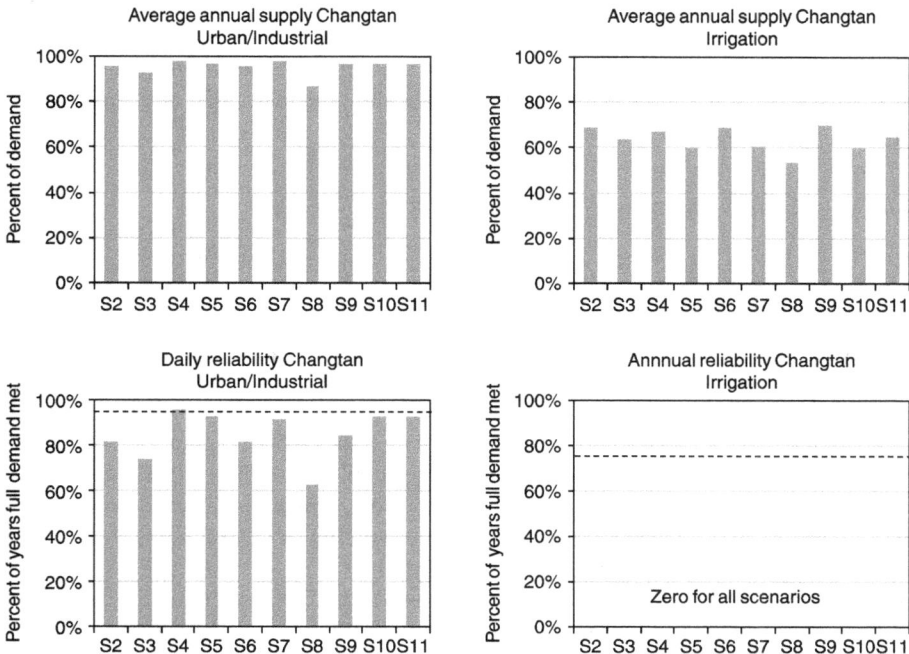

Figure 9. Modelled supply reliabilities for Changtan Reservoir urban/industrial and irrigation users. *Note:* Dashed lines show expected levels of reliability of supply.

by existing infrastructure, subject to some irrigation demand management to bring forward predicted reductions in the sector's needs. All water users on Zhuxi would receive 100% of their demands in 2010, despite there being no reservoir to supplement their supplies. Urban/industrial users from Changtan Reservoir would have a greater daily reliability and average annual supply (relative to demand) than they would in 2020 under the currently planned levels of development. However, on average, irrigators supplied from Changtan Reservoir would only receive 61% of their demands (Figure 9). This is explained by the predicted irrigation demand pattern over the coming decades. Agricultural water use is predicted to increase by about 7% by 2010, then return to existing levels by 2020. Only in the subsequent 10 years are agricultural demands predicted to drop below current levels. Modelling of Scenario S7 indicated that the construction of Zhuxi Reservoir could be delayed until a later date, which could allow for the deferral of capital investment. Also, the final design of the reservoir could be delayed, which in turn could allow for its further optimization (based on the most up-to-date demand information and technology) for environmental and supply security outcomes. Notably, however, additional sources of supply will be needed in order to meet the 2020 demands.

The second less intense development scenario (S8) modelled the construction and operation of all planned development by 2020 except for the pipeline to transfer water from Shisandu Reservoir to Zhuxi Reservoir, i.e. Shisandu Reservoir would supply local demands only. Under this scenario, users of water from Zhuxi Reservoir would receive the same reliability of supply as they would if the pipeline from Shisandu Reservoir was included. The average annual volume of water supplied from Changtan Reservoir would, however, be significantly lower, which would substantially reduce the reliability of supplies received by both urban/industrial and agricultural users (Figure 8). These modelling results suggest that transferring water via a pipeline from Shisandu Reservoir to Zhuxi Reservoir appears necessary to ensure that the predicted 2020 demands for water from Changtan Reservoir can be met.

Implementation of a Sub-optimal Environmental Flows Regime

Scenario S11 was developed to demonstrate a 'trade-off' of increased risk to the maintenance of ecological assets in favour of improved supply reliabilities, by incorporating a sub-optimal environmental flows regime with reduced component frequencies and durations. Scenarios S3, S5 and S10 demonstrated that the provision of the recommended environmental flows in full would impact on the reliability of supply for various users, in particular urban/industrial users supplied from Changtan Reservoir, who would not receive their expected daily reliability of at least 95% (Figure 9). Given that the objective of this scenario (S11) was to maximize supply reliabilities, it also incorporated the modified operation rules for Changtan Reservoir, which had been demonstrated to improve outcomes for urban/industrial users. As such, the results of Scenario S11 should be compared to those of Scenario S5.

Compared to the full recommended environmental flows regime, the sub-optimal regime would not materially change the reliability for urban/industrial users supplied from Changtan Reservoir, but would increase the average annual supply for irrigation from 60% to 65% of these users' demands (Figure 9). This was due to a greater volume of water being transferred from Zhuxi Reservoir to Changtan Reservoir each year. However, this additional water was not sufficient to noticeably improve urban/industrial supply reliability, as increased transfers would not occur during the dry periods, which is the time of greatest shortages. Changing the operational and water sharing rules (e.g. when water is transferred, and also when irrigation water is supplied) could increase urban/industrial supply reliability from Changtan Reservoir (but at a cost to irrigators).

Urban/industrial users supplied from Zhuxi Reservoir would continue to receive 100% of their demands under either the recommended or sub-optimal environmental flows scenarios. The biggest beneficiaries of Scenario S11 would be irrigation users from Zhuxi Reservoir, by receiving 100% supply reliability (Figure 8). Notably, this would be significantly higher than the 75% annual reliability normally expected for agricultural users.

Conclusion

Assessing the environmental flows needs of rivers is usually perceived as a peculiarly 'scientific' task that is undertaken by people who generally are not directly involved in water resources planning decisions. Although this protects the scientific process of determining ecological needs from being 'tainted' by economic and social concerns, it can lead to the situation of planners and stakeholders, not intellectually associated with the original assessment, arbitrarily modifying the recommended flows regime in ways that could make it virtually useless from the perspective of meeting the intended ecological and geomorphological objectives. In this paper, an ecologically-grounded risk assessment methodology was described for generating sub-optimal environmental flows options on the basis of the preferred regime. While some of the principles of the risk assessment may find generic application, others may require modification in particular cases.

The preferred and the sub-optimal environmental flows regimes were incorporated into a water resources allocation planning process, which included a water resources management model capable of modelling the impacts of different operational and supply arrangements on security of supply. The hydrological statistics produced by such models are ideal from the perspective of evaluating the volumes of water available for human use, but they may not be useful for assessing the likelihood of environmental objectives being met. This is because the ecological elements of the aquatic environment can respond to fairly subtle and complex aspects of the flow regime. A new method was presented here for calculating the degree of compliance of a flow series with a specified environmental flows regime. The onus is on the scientific panel that is charged with assessing the flow needs of the river to specify the required regime in sufficient detail such that all aspects of compliance can be objectively assessed. When the compliance method was applied to four river reaches in the Jiaojiang Basin, it unambiguously discriminated the aspects of the flow regime that were deficient from the perspective of protecting ecological assets with a low level of risk.

The comprehensive water resources planning process described in this paper was applied to the Jiaojiang Basin in Zhejiang Province. The main conclusions arising from that investigation were:

- Generally, the 2010 demands could be met from within existing supplies (i.e. without the development of Zhuxi Reservoir), subject to the irrigation demands from Changtan Reservoir being reduced to the 2020 demand levels. While enabling the deferral of some capital investment and the time for further design work, Zhuxi Reservoir will still be required by 2020.

- The transfer of water from Shisandu Reservoir to Zhuxi Reservoir would be unnecessary in order to meet the 2020 demands on Zhuxi Reservoir. However, the transfer would be necessary in order to meet the 2020 demands from Changtan Reservoir.
- The currently proposed arrangements for 2020 will not achieve acceptable levels of supply reliability for users supplied from Changtan Reservoir. This was a significant finding from the work, as it showed that the supply objectives for the basin would not be met regardless of the adoption of any environmental flows.
- Lowering the restriction threshold for urban/industrial water supplies from Changtan Reservoir would significantly improve the supply reliability (both daily and annual) for these users with only a small reduction in annual reliability for irrigators.
- The planned 2020 developments and infrastructure operating rules would have a major impact on environmentally important flows in Zhuxi, as well as impacts to lesser degrees further down the system to the Jiaojiang estuary.
- Adopting the recommended environmental flows would play a significant role in protecting the ecological assets of the river system, with no negative consequence for urban/industrial supplies from Zhuxi Reservoir. However, there would be various reductions in the reliability of irrigation supplies from Zhuxi Reservoir and all supplies from Changtan Reservoir.
- The environmental flows recommended would significantly reduce the annual volume of water transferred from Zhuxi Reservoir to Changtan Reservoir.
- Adoption of the environmental flows recommendations would require the operational rules for Zhuxi Reservoir to be redesigned.
- Alternative, sub-optimal environmental flows would provide higher irrigation supply reliability than under the recommended flows regime, but not make much difference to urban/industrial users. These sub-optimal rules would increase the risk of not maintaining the health of the identified ecological assets.
- It appears that there are options available that offer advantages, both in terms of supply reliability and environmental flows compliance, compared to the arrangements that are currently planned for 2020.

This paper did not recommend a particular scenario to be incorporated into a water resources allocation plan for Taizhou. This is a decision that should be made by the appropriate level of government following further consideration and analysis of the implications of various scenarios (including additional scenarios not contemplated in this investigation) by appropriate stakeholders. Nor did this paper recommend a threshold upper limit of development beyond which the risk to river health was considered intolerable, although this could potentially be achieved through further model development, data collection and consultation with stakeholders and managers. Regardless, river managers and water resource planners need to be aware that there is a point at which the impacts of development on river health will become obvious even to the casual observer, let alone those dependent on fisheries for a livelihood who may have been suffering reduced yields for some time. At that point, it may be an extremely difficult and expensive exercise to restore river health. Even where that threshold is unknown, resource managers should be aware of the risk, monitor trends in relevant indicators and apply environmental flows rules that minimize the risk to important ecological assets.

Acknowledgements

This paper is the result of a project undertaken under the auspices of the Australian Department of the Environment, Water, Heritage and the Arts and the Chinese Ministry of Water Resources, with funding provided by AusAID, the Australian Agency for International Development. The Department of Water Resources, Zhejiang Province, supported the field component.

References

Brown, C. A. & King, J. M. (2003) Environmental flows: Concepts and methods, in: R. Davis & R. Hirji (Eds) *Water Resources and Environment Technical Note C1* (Washington, DC: The World Bank).

Cai, X. & Rosegrant, M. W. (2004) Optional water development strategies for the Yellow River Basin: Balancing agricultural and ecological water demands, *Water Resources Research*, 40, W08S04, doi: 10.1029/2003WR002488.

Department of Land and Water Conservation (1998) *Integrated Quantity Quality Model (IQQM) User Manual*, Version 6.33 (Parramatta, NSW: New South Wales Government).

Gippel, C. J. & Stewardson, M. J. (1995) Development of an environmental flow management strategy for the Thomson River, Victoria, Australia, *Regulated Rivers: Research & Management*, 10, pp. 121–135.

Gippel, C. J., Bond, N. R., James, C. & Wang, X. (2009) An asset-based, holistic, environmental flows assessment approach, *International Journal of Water Resources Development*, 25(2), pp. 301–330.

Harman, C. & Stewardson, M. (2005) Optimizing dam release rules to meet environmental flow targets, *River Research and Applications*, 21, pp. 113–129.

King, J. M., Brown, C. A. & Sabet, H. (2003) A scenario-based holistic approach to environmental flow assessments for rivers, *River Research and Applications*, 19(5–6), pp. 619–639.

Palmer, R. N. & Snyder, R. M. (1985) Effects of instream flow requirements on water supply reliability, *Water Resources Research*, 21(4), pp. 439–446.

Poff, N. L., Allan, J. D., Bain, M. B., Karr, J. R., Prestegaard, K. L., Richter, B. D., Sparks, R. E. & Stromberg, J. C. (1997) The natural flow regime, a paradigm for river conservation and restoration, *BioScience*, 47, pp. 769–784.

Richter, B. D., Baumgartner, J. V., Wigington, R. & Braun, D. P. (1997) How much water does a river need? *Freshwater Biology*, 37, pp. 231–249.

Suen, J-P. & Eheart, J. W. (2006) Reservoir management to balance ecosystem and human needs: Incorporating the paradigm of the ecological flow regime, *Water Resources Research*, 42(3), W03417, 10.1029/2005WR004314.

Taizhou Water Resources Bureau (2004) *Comprehensive Plan for Water Resources in the Taizhou Prefecture*, Translated by Water Entitlements and Trading Project (Taizhou, Zhejiang, P.R. China: Taizhou Water Resources Bureau).

Vogel, R. M., Sieber, J., Archfield, S. A., Smith, M. P., Apse, C. D. & Huber-Lee, A. (2007) Relations among storage, yield, and instream flow, *Water Resources Research*, 43, W05403, doi:10.1029/2006WR005226.

WET (2007) *Water Entitlements and Trading Project (WET Phase 2) Final Report* December 2007 [in English and Chinese] (Beijing: Ministry of Water Resources, People's Republic of China and Canberra: Department of the Environment, Water, Heritage and the Arts, Australian Government). Available at: http://www.environment.gov.au/water/action/international/wet2.html

Wollmuth, J. C. & Eheart, J. W. (2000) Surface water withdrawal allocation and trading systems for traditionally riparian areas, *Journal of the American Water Resources Association*, 36(2), pp. 293–303.

A Harmonious Water Rights Allocation Model for Shiyang River Basin, Gansu Province, China

ZHONGJING WANG, HANG ZHENG & XUEFENG WANG

ABSTRACT *This paper summarizes water rights allocation principles based on the experience of international and domestic water rights allocation, and presents a water rights allocation model based on the principles of security, sustainability, fairness and efficiency. Applying the model to the Shiyang River Basin in Gansu Province, China, surface water and the groundwater rights in the basin are defined and allocated for current and future years. A comparison between allocation results from this study and water allocation plans set out in the "Focus Restoration Plan of Shiyang River Basin" demonstrates that water allocation based on current levels and patterns of water use is relatively straightforward, but that defining and allocating rights according to future demands and management needs is more uncertain. Nevertheless, to address the serious problem of water resource over-exploitation in the Shiyang River Basin, an initial water rights allocation based on future projections of planned water use is proposed. This will help support water conservation efforts in areas such as Minqin County, a downstream area of the Shiyang River where water resources degradation is particularly severe.*

Introduction

In recent decades, much attention has been focused on water rights systems for supporting improvements in water resources management in China (Gao, 2006). An initial allocation of water rights to determine annual water use caps for different users in a reasonable and transparent way underpins better water resources management. Water conservation, environmental protection, rational development and utilization of water resources, conflict resolution and the development of water markets all depend on the definition and allocation of clear rights. Developing and implementing a modern system of water rights across China is a complex task, however. In this paper, a model is presented that can help water planners meet this challenge, based on a set of core principles.

Initial Water Rights Allocation Principles

Allocation principles form the basis for water rights allocation, setting out clearly the multiple objectives of water utilization. A set of reasonable and acceptable principles are also needed to ensure political and public acceptability. There are a number of water rights allocation principles applied in different counties, including those of riparian ownership, prior appropriation and public rights. International experience highlights how these principles have evolved according to local context (Ge, 2002). In regions with relatively abundant water such as Europe and eastern America, riparian principles have dominated. Conversely, in regions of relative scarcity such as western America, the prior appropriation doctrine has dominated, supplemented by riparian rights. In Japan, both 'upstream priority' and 'first in time, first in right' principles have been implemented together. In China, water allocation principles have also evolved according to changing political and economic priorities. International comparisons provide some lessons, but there is no simple blueprint for China (Ge, 2002).

The adoption of water rights allocation principles depends on history, the objectives of water resources management and the actual conditions of water resources in a river basin. In order to achieve the rational and efficient use of water, it is necessary to select principles according to 'facts on the ground'. In China, experts such as Wang (2001), Liu (2003), Lin (2002), Ge (2003) and Ge (2004) have put forward different water rights allocation principles. This paper seeks to re-classify and compare them, and aims to determine a set of principles that can guide the initial allocation of water rights in China, focusing on a 'Basic Water Demand Guarantee', 'Sustainable Development' and 'Fairness and Efficiency'.

Basic Water Demand Guarantee

Basic water demand is defined as the water needed for basic living, basic ecology, basic economics and basic crop production. These universal entitlements are needed to support human existence, key environmental services and food security. Generally, in water allocation, basic water demands should receive the highest priority, be reserved in advance and be satisfied first.

Sustainable Development

There is now a common consensus that water resource allocation and development should be informed by an understanding of system sustainability. In this paper, the sustainable development principle focusses on 'ecological' water demand. This, in turn, is divided into 'basic' and 'exclusive' demands. Basic demand refers to the water needed to maintain the basic (out-of-stream) ecology of a basin and to ensure safe water quality and flows, and should be satisfied under the Basic Water Demand Guarantee principle outlined above. 'Exclusive' ecological water refers to the water needed to restore and develop the ecology of a basin—both within and outside the watercourse. This is determined by the objectives set for ecological restoration and available water in the river basin.

Fairness

There are several important factors that affect people's perception of fairness with respect to water rights allocation. These include prior use and customary rights, population served, the irrigation area, the contribution of water to livelihoods and production, water shortages experienced and the future needs of different stakeholders, including environmental needs.

For water use to be considered 'fair', all of these factors need to be considered and balanced against each other.

Efficiency

The efficiency principle relates to the economic efficiency of water use. Efficiency criteria include the productivity of water use, in terms of (for example) revenue or output per unit of water consumptively used.

In most cases, there are contradictions between fairness and efficiency in water allocation (Wang, 2006). Fairness requires that the water allocated to stakeholders is relatively equal or proportional, while efficiency requires that those stakeholders with higher water use efficiency and greatest 'income or revenue per drop' should get more water. However, it is difficult to satisfy both at the same time. From a legal and ethical point of view, fairness should be the main priority in initial water rights allocation, with efficiency then addressed through water trading. Fairness embodies the factors listed above, and is approached most readily through a consideration of current patterns of water use in relation to these factors.

Integrated Surface Water and Groundwater Management

An initial allocation of water rights needs to consider interactions between surface water and groundwater, as these systems are often inter-related in the water cycle. For example, groundwater can provide base flow to rivers, and irrigation returns can provide significant groundwater recharge.

A Harmonious Rights Allocation Model

In this section, an optimization model is presented that can assist in the initial definition and allocation of water rights for different counties within the Shiyang River Basin, based on a 'Satisfactory Function'. This is a quantitative method for considering the allocation principles and factors described above. A 'Genetic Algorithm' (GA) is used to compute this multi-objective, non-linear model. No matter how principles are considered, trade-offs between competing needs and objectives are unavoidable. The multi-objective optimization model weights and then combines competing needs, with the objective of maximizing combined values, subject to certain constraints.

The Objective Function

The objective function is shown as equation 1.

$$\max S = \omega_1 \cdot RBS + \omega_2 \cdot RES + \omega_3 \cdot RFS + \omega_4 \cdot RHS \tag{1}$$

Where, *RBS*, *RES*, *RFS* and *RHS* are satisfactory functions for basic food security, the exclusive ecological water demand guarantee, fairness and efficiency principles, respectively. ω_j ($j = 1, 2, \ldots, 4$) describes the weighting used for each. Satisfactory functions are available for both surface water right allocation and the ground water allocation, shown below.

RBS: Basic food security, i.e. crop production water demand guarantee.

$$RBS_i = \begin{cases} \frac{WR_i}{W_{ABi}} & WR_i < W_{ABi} \\ 1 & WR_i \geq W_{ABi} \end{cases} \quad i = 1, 2, \ldots, n \tag{2}$$

$$RBS = \begin{cases} 1 & \min(RBS_i) = 1 \\ \frac{\min(RBS_i - 0.95)}{1 - 0.95} & 0.95 < \min(RBS_i) < 1 \\ 0 & \min(RBS_i) \leq 0.95 \end{cases} \tag{3}$$

Where, W_{ABi} is the basic water demand for crop production of the county i; WR_i is the sum of the surface and ground water rights allocated for the economic use of county i, which does not include the allocated basic living water and ecological water.

RES: Ecological water demand guarantee.

$$RES = \begin{cases} \frac{W'_{ED}}{W_{ED}} & W'_{ED} < W_{ED} \\ 1 & W'_{ED} \geq W_{ED} \end{cases} \tag{4}$$

$$W_{EB} + W_{ED} = W_E \tag{5}$$

Where, W_{ED} is the total exclusive ecological water demand of all counties; W_{EB} is the total basic ecological water demand of all counties; W'_{ED} is the total water (both from surface and groundwater sources) allocated for the exclusive ecological water demand of all counties; and W_E is the total ecological water demand in the river basin.

RFS: Fairness. The factors that need to be considered under the fairness principle include appropriation, population, area, water contribution and water shortage. An integrated function for fairness is set up by combining functions and their weights. The function of *RFS* shown in equation 6 is available for both surface water allocation and groundwater allocation.

$$RFS = \beta_1 \cdot RF_1 + \beta_2 \cdot RF_2 + \beta_3 \cdot RF_3 + \beta_4 \cdot RF_4 + \beta_5 \cdot RF_5 \tag{6}$$

Where, RF_1, RF_2, RF_3, RF_4 and RF_5 are the satisfactory functions for 'prior appropriation', 'prior population', 'prior area', 'prior water contribution' and 'prior water shortage', respectively, and $\beta_j (j = 1, 2, \ldots, 5)$ is the weight coefficient of each factor. Separate equations for each of these functions were developed, allowing, amongst other things, for the surface and groundwater allocations, the irrigated area of each county and the total water resource in the basin.

RHS: The satisfactory function of the efficiency principle.

$$RHS = \frac{\sum_{i=1}^{n} \frac{WR_i \cdot GDP_i}{WO_i} - W_U \cdot \min\left(\frac{GDP_i}{WO_i}\right)}{W_U \cdot \max\left(\frac{GDP_i}{WO_i}\right) - W_U \cdot \min\left(\frac{GDP_i}{WO_i}\right)} \quad i = 1, 2, \ldots, n \tag{7}$$

Where, GDP_i is the GDP production of county i in the base year; WO_i is the surface and groundwater economic water use of county i in the base year; W_U is the total economic water use (surface and ground) of the whole river basin.

Water Balance Constraint

$$WR_i = WSR_i + WGR_i \tag{8}$$

$$\sum_{i=1}^{n} WR_i = W_R \tag{9}$$

$$W'_{LBi} = WS'_{LBi} + WG'_{LBi}; \quad W'_{EBi} = WS'_{EBi} + WG'_{EBi}; \quad W'_{EDi} = WS'_{EDi} + WG'_{EDi} \tag{10}$$

$$\sum_{i=1}^{n} W'_{LBi} = W'_{LB}; \quad \sum_{i=1}^{n} W'_{EBi} = W'_{EB}; \quad \sum_{i=1}^{n} W'_{EDi} = W'_{ED} \tag{11}$$

$$W_R + W'_{LB} + W'_{EB} + W'_{ED} = W_T \tag{12}$$

Where, W_R is the allocated economic water right of all counties (not including the allocated ecological and basic living water); WSR_i is the surface water rights allocated for the economic use of county i; WGR_i is the ground water rights allocated for the economic use of county i; W_T is the total water resource of a river basin; n is the number of counties; WS'_{LBi} is the allocated surface water rights for basic living for county i; WG'_{LBi} is the allocated groundwater rights for basic living for county i; WS'_{EBi} is the allocated surface water rights for basic ecology for county i; WG'_{EBi} is the allocated groundwater rights for basic ecology for county i; WS'_{EDi} is the allocated surface water rights for the exclusive ecology for county i; WG'_{EDi} is the allocated groundwater rights for exclusive ecology for county i; W'_{LBi} is the allocated basic living water rights for county i; W'_{EBi} is the allocated basic ecological water rights for county i; W'_{EDi} is the allocated exclusive ecological water rights for county i; W'_{LB} is the total water (from both surface water and groundwater sources) allocated for the basic living water demand of all counties; W'_{EB} is the total water (from both surface water and groundwater sources) allocated for the basic ecological water demand of all counties; W'_{ED} is the total water (from both surface water and groundwater sources) allocated for the exclusive ecological water demand of all counties.

Key Issues in Water Rights Allocation

There are several key issues that need to be considered further in water rights allocation. These are described below.

Basic Water Allocation

Given the priority attached to meeting basic needs, the water rights needed to satisfy basic living conditions (i.e. domestic water needs) and basic ecology (out-of-stream environmental water use) are allocated first, before exclusive ecological and economic water rights are allocated. By first subtracting the water required to meet basic living and ecological rights from the total water resources of the basin, we derive the available water resources for meeting exclusive ecological needs and economic uses. This water is allocated to counties using the optimization model and functions described previously through equations 1 to 12.

Currently, basic water demand is satisfied from both surface and groundwater sources. Notably, basic living and ecological water demands are much less than those for economic uses. To respect and maintain current patterns of (domestic) water use in the basin, rights to water for basic living and ecology are allocated from both surface and groundwater.

Ecological Water Allocation

The allocation of ecological water rights in this study is considered in two phases. Basic ecological water rights for maintaining basic out-of-stream ecology (e.g. for watering forests) in the river basin are allocated first. Exclusive ecological water rights for restoring and developing the ecological system are then allocated together with economic water rights. To meet the broad needs and 'public good' attributes of the ecological system, ecological water rights are allocated across the basin first, rather than to each county. The total ecological water rights of the basin can then be allocated to individual counties, based on their actual ecological water demands and ecological restoration objectives.

Current Water Use and Future Scenarios

Two alternative sets of water use data can be used for the water allocation model. The first option is to use current data on water use. The second is to use planned or forecast data on water use. In the sections below, both options are explored and results evaluated, using data for the year 2000 (the 'current' baseline) and projected data for 2020.

Surface Water and Groundwater Allocation

Both surface water and groundwater rights are allocated to counties in this study. As a general rule, it is proposed that surface water should be used prior to groundwater, as groundwater resources in the basin are already over-exploited, or nearing the point at which they will become so. However, it is recognized that groundwater meets many of the basic living and ecological needs of counties, and that demand is widely dispersed across the basin. Hence, it is not necessary or feasible to satisfy these demands from surface sources first. The model therefore assumes that existing shares or proportions of water drawn from surface and groundwater sources are maintained in the future with respect to basic living and ecological water demands.

In contrast, economic water demand, which is both larger and more spatially concentrated, can be satisfied predominately from surface water. However, where the economic water demand cannot be satisfied from surface water sources, groundwater will be allocated according to the surface water deficit and the current economic groundwater use of counties. In other words, original surface and groundwater allocations for economic water uses change over time. Some groundwater demand is therefore met through the allocation of surface water rights. This economic water allocation method is designed to prevent groundwater overdraft, and is particularly relevant in those river basins where groundwater is non-renewable, and/or where intensive use of groundwater threatens sustainability.

Water Rights Allocation in Shiyang River Basin

The Shiyang River Basin is an inland watershed with the largest population, the most developed economy, and the highest level of water resources development in Gansu Province, in the northwest of China. The location of the river basin is shown in Figure 1. The river basin covers an area of $41\,600\,km^2$, with five administrative units at county level: Gulang County, Liangzhou District, Minqin County, Yongchang County and Jinchuan District (Figure 2). The total renewable water resources of the basin are estimated at 1.66 billion m^3, including an annual average surface water resource of 1.56 billion m^3, and 99 million m^3 groundwater resources. Total water withdrawals in the river basin were estimated at 1.71 billion m^3 in 2000 and hence water resources are already over-exploited.

Water resource over-exploitation has caused serious ecological problems and constrained social and economic development in the basin (Gao, 2006). Ecological deterioration in Minqin County has been particularly severe, creating 'ecological refugees' as people have been forced to abandon homes and livelihoods. To ease the water resource shortage and address the problem of ecological deterioration in the basin, the Water Resource Department (WRD) and the Development & Reform Commission (DRC) of Gansu Province developed a "Restoration Plan for Shiyang River Basin in the near future" (or the Restoration Plan). This was completed in 2005 and approved by the State Council in 2007 (Gansu WRD & DRC, 2005).

The sections below summarize how the model can allocate the total water resources of Shiyang River Basin to each administrative unit according to the principles outlined above, to address these problems.

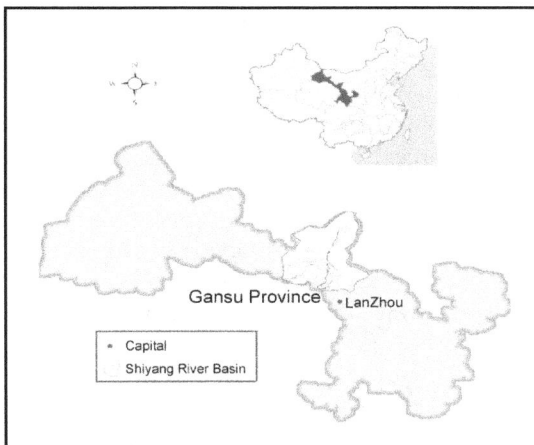

Figure 1. Location of Gansu Province and the Shiyang River Basin. *Source:* Gansu WRD and DRC (2005).

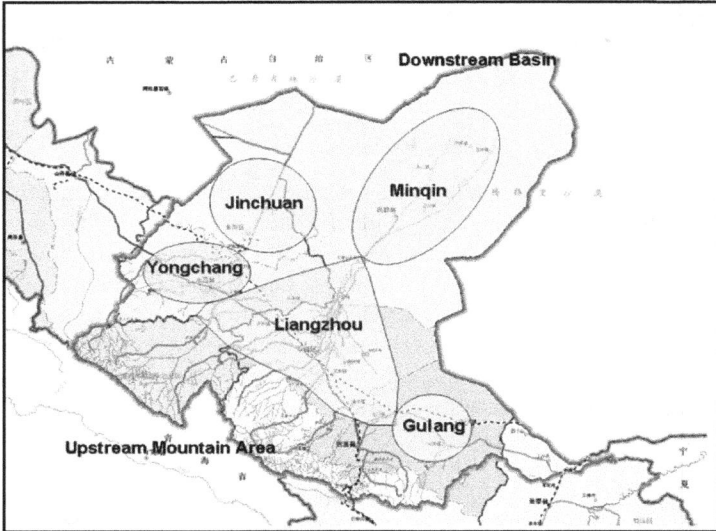

Figure 2. Administrative zones of the Shiyang River Basin. *Source:* Gansu WRD and DRC (2005).

Water Allocation Based on Current Water Use

Based on data for the year 2000, the current ecological water demand (i.e. the water needed to maintain the essential existing out-of-stream ecological systems) is estimated at 0.1 billion m^3. This figure is used in the model. This and other data for the year 2000 used for the model are presented in Table 1.

Surface and groundwater resources are both allocated based on current water use (year 2000 data). No exclusive ecological water demands or reserves are allocated because resources are already over-exploited. Allocation results from the model—for living, ecological and economic uses—are presented in Table 2, for comparison against actual water use at that time. The basic living and ecological water rights allocated by the model (Table 2) are equal to the basic living and ecological water uses in 2000 (Table 1). Hence, basic living and ecological water demands from both surface water and groundwater sources are completely satisfied by the water rights allocated amongst the counties.

For economic water uses, however, the allocations for surface water sources granted under the model are all larger than the actual use from those sources in 2000. Conversely, economic allocations from groundwater sources are much lower as the model operates to reduce groundwater overdraft. (Renewable groundwater in the basin is estimated at 99 million m^3, but current abstraction is estimated at 923.9 million m^3). Hence the model allocates more surface water for economic use and reduces groundwater allocations to balance demand with renewable supply. These differences are illustrated in Figure 3.

Water Allocation Based on Future Water Demand

Water rights can also be allocated according to projected development needs and plans. The Shiyang River Basin Restoration Plan includes water allocations for each county for 2020, as well as expected water usage at that time (often lower than the volume allocated).

Table 1. Actual water withdrawals in the Shiyang River Basin for the year 2000

		Gulang	Liangzhou	Minqin	Jinchuan	Yongchang	Total
Surface water consumption							
Basic water consumption (10^6 m³)	For living (WLB)	2.7	27.6	0.0	9.4	4.3	43.9
	For ecology (WEB)	2.1	19.0	50.0	0.8	5.0	76.9
	For crop production (WAB)	41.3	291.7	170.0	77.6	53.4	634.0
	For economics (WSO)	41.1	337.4	60.9	103.8	124.8	668.0
Total water consumption (10^6 m³)		45.9	384.0	110.9	113.9	134.1	788.8
Groundwater consumption							
Basic water consumption (10^6 m³)	For living (WLB)	2.6	17.0	10.9	0.8	5.0	36.2
	For ecology (WEB)	0.1	9.7	12.5	0.0	3.5	25.8
	For crop production (WAB)	0.0	0.0	0.0	0.0	0.0	0.0
	For economics (WGO)	9.4	301.8	391.9	67.8	91.1	861.9
Total water consumption (10^6 m³)		12.0	328.5	415.3	68.5	99.6	923.9
Total surface and groundwater consumption (WO) (10^6 m³)		58.0	712.4	526.2	182.4	233.7	1712.7
Population (10^3)		247.6	978.6	302.1	258.4	213.4	2000.1
Area (10^3 mu*)		320.5	1746.1	1065.2	204.6	1141.2	4477.6
GDP (10^6 RMB)		601.0	5361.0	1204.0	4517.0	1112.0	12795.0

*1 mu = 0.067 ha.
Source: Gansu WRD & DRC (2005).

Table 2. Proposed (modelled) allocations versus actual water use for the year 2000

Water Rights & Water Use (10^6 m^3)		Gulang	Liangzhou	Minqin	Jinchuan	Yongchang	Total
Proposed allocations (surface) 2000	Basic living	2.7	27.6	0.0	9.4	4.3	44.0
	Basic ecology	2.1	19.0	50.0	0.8	5.0	76.9
	Economic	47.9	604.2	424.6	161.1	202.4	1440.2
	Total	52.7	650.8	474.6	171.2	211.7	1561.0
Proposed allocations (groundwater) 2000	Basic living	2.6	17.0	10.9	0.8	5.0	36.2
	Basic ecology	0.1	9.7	12.5	0.0	3.5	25.8
	Economic	0.4	13.5	16.2	3.0	3.8	37.0
	Total	3.1	40.3	39.5	3.8	12.3	99.0
Total of surface and groundwater allocations		55.8	691.0	514.2	175.0	224.0	1660.0
Actual water use (surface) 2000	Basic living	2.7	27.6	0.0	9.4	4.3	44.0
	Basic ecology	2.1	19.0	50.0	0.8	5.0	76.9
	Economic	41.1	337.4	61.0	103.8	124.8	668.0
	Total	45.9	384.0	111.0	113.9	134.1	788.8
Actual water use (groundwater) 2000	Basic living	2.6	17.0	10.9	0.8	5.0	36.2
	Basic ecology	0.1	9.7	12.5	0.0	3.5	25.8
	Economic	9.4	301.8	391.9	67.8	91.1	861.9
	Total	12.1	328.5	415.3	68.5	99.6	923.9
Actual total surface and groundwater use		58.0	712.4	526.2	182.4	233.7	1712.7

Source: Guansu WRD & DRC (2005).

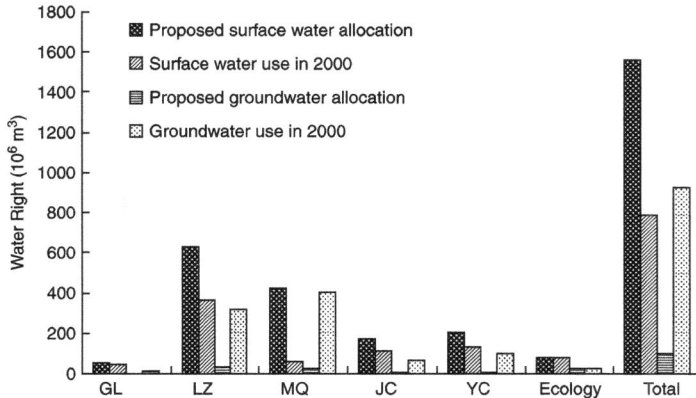

Figure 3. Actual water use in 2000 and proposed (modelled) allocations. *Note:* GL – Gulang County; LZ – Liangzhou District; MQ – Minqin County; JC – Jinchuan District; YC – Yongchang County. *Source for actual usage*: Guansu WRD & DRC (2005).

Data for 2020 extracted from the plan are shown in Table 3. The plan anticipates a reduction in total water use in the basin to 1.36 billion m³ by 2020, 0.35 billion m³ less than total water use in 2000 and less than total surface water available in the basin. This includes a reduction in surface water use to 749.7 million m³ by 2020 (a little less than the year 2000 figure of 788.8 million m³), and a significant reduction in groundwater abstraction to 612.6 million m³, considerably less than the figure for groundwater abstraction in 2000 (923.9 million m³).

Data from Table 3 are then used in the model to derive alternate surface and groundwater allocations for basic living, ecology and economic purposes. As a result of economic development and water saving in the future, the basic ecological water demand in 2020 is estimated at 0.8 billion m³, with no additional allocation made for exclusive ecological demands or reserved water. Surface and groundwater are both allocated, based on future water use in the basin. Allocation results from the model are shown in Table 4 and Figure 4. Basic living and ecological water demands in 2020 from surface and groundwater sources are fully met by the rights assigned to the different counties. Although groundwater use in 2020 is much reduced, it still significantly exceeds sustainable yield within the basin. The surface water rights allocated by the model are increased to compensate for the reduction in groundwater withdrawals, and hence surface water allocations are greater than those proposed in the Restoration Plan for 2020.

Comparison and Discussion

To validate the allocation results above and analyse the impacts of different factors and weights in the model, model outputs were compared with the proposed allocations and expected usage in the Restoration Plan and the actual water use situation in 2000. These results are shown in Figure 5, 6 and 7.

The Restoration Plan allocates 61 million m³ of water for basic ecology and reserves 73 million m³, out of a total water use cap of 1.6 billion m³. The results show some differences between current water consumption, water allocations based on water use in 2000 and 2020, water allocation based on the Restoration Plan and water consumption

Table 3. Planned water withdrawals in the Shiyang River Basin for 2020

		Gulang	Liangzhou	Minqin	Jinchuan	Yongchang	Total
Surface water consumption (10^6 m^3)							
Basic water consumption	For living (WLB)	6.6	45.5	0.0	14.0	6.4	72.4
	For ecology (WEB)	6.4	26.2	0.0	5.0	7.1	44.7
	For crop production (WAB)	68.7	211.2	66.7	31.4	34.8	412.8
	For economics (WSO)	76.9	302.4	25.6	126.1	101.6	632.6
Total water consumption (10^6 m^3)		89.9	374.1	25.6	145.1	115.1	749.7
Groundwater consumption (10^6 m^3)							
Basic water consumption	For living (WLB)	2.3	15.7	16.8	0.6	9.8	45.2
	For ecology (WEB)	0.3	13.4	21.0	0.0	5.0	39.7
	For crop production (WAB)	0.0	0.0	0.0	0.0	0.0	0.0
	For economics (WGO)	15.6	246.5	166.3	27.4	72.0	527.8
Total water consumption (10^6 m^3)		18.1	275.6	204.0	28.0	86.8	612.6
Total surface and groundwater consumption (WO) (10^6 m^3)		108.1	649.8	229.6	173.0	201.9	1362.4
Population (10^3)		185.1	1084.5	336.5	228.2	276.2	2110.5
Area (10^3 mu*)		298.4	1556.3	625.3	157.5	505.4	3142.9
GDP (10^6 RMB)		3028.0	26886.0	3220.0	34370.0	10889.0	78393.0

* 1 mu = 0.067 ha

Source: Gansu WRD & DRC (2005).

Table 4. Modelled allocations and anticipated water usage for 2020

Water Rights & Water Use (10^6 m^3)		Gulang	Liangzhou	Minqin	Jinchuan	Yongchang	Total
Water allocations (surface) 2020	Basic living	6.6	45.5	0.0	14.0	6.4	72.4
	Basic ecology	6.4	26.2	0.0	5.0	7.1	44.7
	Economic	116.3	681.4	241.1	190.0	215.2	1443.9
	Total	129.2	753.1	241.1	209.0	228.7	1561.0
Water allocation (ground) 2020	Basic living	2.3	15.7	16.8	0.6	9.8	45.2
	Basic ecology	0.3	13.4	21.0	0.0	5.0	39.7
	Economic	1.1	6.7	2.3	1.9	2.1	14.1
	Total	3.7	35.8	40.1	2.5	16.9	99.0
Sum of surface and groundwater rights		132.9	788.9	281.2	211.5	245.6	1660.0
Anticipated water usage (surface) 2020	Basic living	6.6	45.5	0.0	14.0	6.4	72.4
	Basic ecology	6.4	26.2	0.0	5.0	7.1	44.7
	Economic	76.9	302.4	25.6	126.1	101.6	632.6
	Total	89.9	374.1	25.6	145.1	115.1	749.7
Anticipated water usage (ground)2020	Basic living	2.3	15.7	16.8	0.6	9.8	45.2
	Basic ecology	0.3	13.4	21.0	0.0	5.0	39.7
	Economic	15.6	246.5	166.3	27.4	72.0	527.8
	Total	18.1	275.6	204.0	28.0	86.8	612.6
Sum of surface and groundwater use		108.1	649.8	229.6	173.0	201.9	1362.4

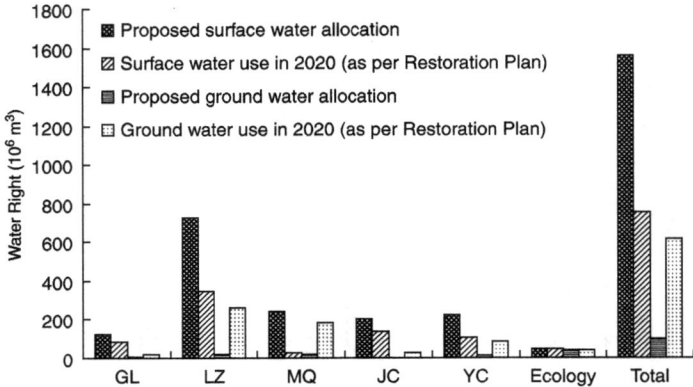

Figure 4. Proposed (modelled) allocations and expected water use in 2020. *Note:* GL – Gulang County; LZ – Liangzhou District; MQ – Minqin County; JC – Jinchuan District; YC – Yongchang County.

in 2020 based on the model. There is a large difference between current water use (in 2000) and the expected water usage in 2020, as per the Restoration Plan. However, model allocations for 2020 are much closer to the expected water use in the Plan, particularly for Minqin County. This is because the Restoration Plan is based on water saving and efficiency assumptions.

To ensure sustainable use of water resources, there will need to be a significant reduction in water use in Minqin County. Therefore, a comparison of model results for 2000 and 2020 show a significant reduction in allocations to this county, with correspondingly more water allocated to the upper and middle reaches of the basin in the future, illustrated in Figure 7. Hence water saving pressure in the basin is more focused in the lower reaches, and especially in Minqin County, as shown in Figure 8. The water saving requirement of Minqin County is 203.59 million m^3 from its current actual water use (2000), to achieve the modelled allocation result for 2020. There will need to be a

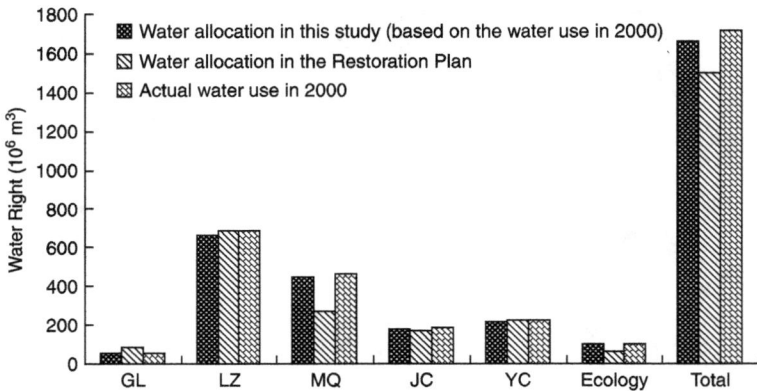

Figure 5. Comparison of modelled allocations (rights) for 2000, actual water use in 2000 and allocations granted under the Restoration Plan. *Note:* GL – Gulang County; LZ – Liangzhou District; MQ – Minqin County; JC – Jinchuan District; YC – Yongchang County.

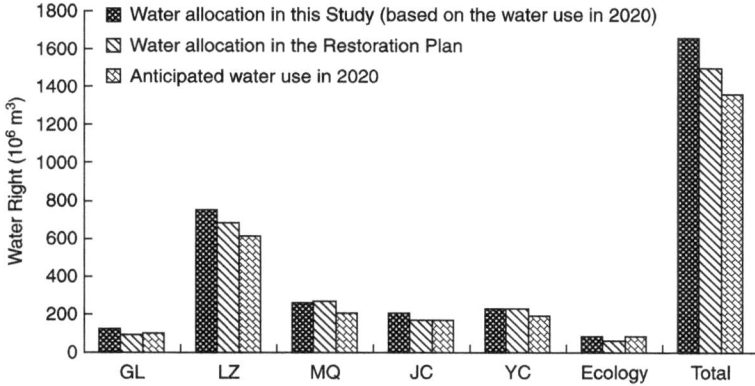

Figure 6. Comparison of modelled allocations (rights) for 2020, allocations granted under the Restoration Plan and anticipated total water use in 2020. *Note:* GL – Gulang County; LZ – Liangzhou District; MQ – Minqin County; JC – Jinchuan District; YC – Yongchang County.

major reduction in water usage if the county is to comply with its proposed water allocation level. In view of expected progress on water saving, an allocation of rights based on future water use (after saving) is more appropriate for the Shiyang River Basin. In particular, the potential for water saving in Minqin County is substantial in comparison with upstream areas because its economy is still predominantly agricultural.

Conclusions

This study has presented an optimal water rights allocation model based on harmonious allocation principles, and applied it in the Shiyang River Basin. There are three main advantages over the rights allocation process proposed by Wang *et al.* (2006). First, the principle of fairness discussed in this study is important, especially in heavily exploited

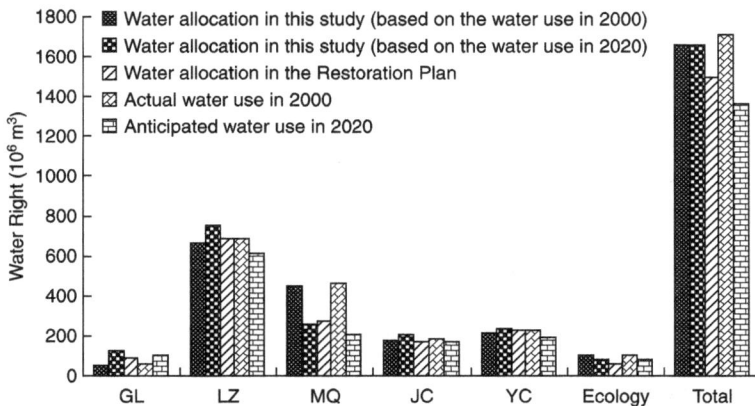

Figure 7. Comparison of modelled allocations, allocations in Restoration Plan, actual total water use in 2000 and expected total use for 2020. *Note:* GL – Gulang County; LZ – Liangzhou District; MQ – Minqin County; JC – Jinchuan District; YC – Yongchang County.

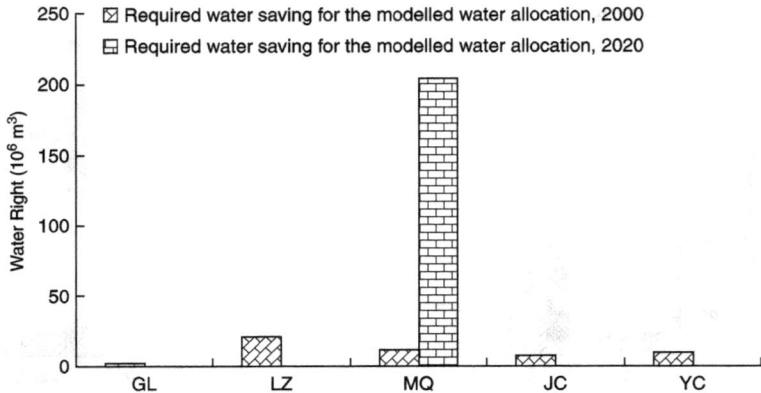

Figure 8. Water savings required for model allocations in 2000 and 2020. *Note:* GL – Gulang County; LZ – Liangzhou District; MQ – Minqin County; JC – Jinchuan District; YC – Yongchang County. *Note:* "Required water saving" means how much water should be saved, to achieve the modelled water allocations from the current actual water use situation (2000).

river basins where water is scarce. So in the allocation model, we adopt a function to represent the fair use principle, based on existing uses, population, irrigated area, the contribution water makes to the economy and the water scarcity experienced by different counties, rather than basing rights allocation on prior appropriation only (Wang *et al.*, 2006). Second, both surface water and groundwater resources are considered together and allocated. And finally, the study illustrates how allocation outcomes can be compared, based on comparisons between current use and future projections, and model outputs for current and future allocations.

In 2000 and 2020, existing and planned (respectively) levels of groundwater use in the Shiyang River Basin exceed sustainable limits, yet surface water use in both years is less than available supply. The model described in this paper addresses this problem by adjusting surface and groundwater withdrawals, allocating more surface water and less groundwater for economic uses. To meet basic living and ecological needs, surface water and groundwater are allocated according to the proportions used in the base year, in this case 2000. For economic water rights, however, surface water is allocated first, whilst meeting dispersed groundwater demands as far as possible. The surface water rights allocated to counties by the model are larger than existing (year 2000) and planned (2020) allocations, whilst the groundwater rights allocated are lower. In order to implement modelled allocations, the infrastructure for the delivery of surface water allocations needs to improve, with concomitant improvements in storage capacity to ensure supply at times of peak demand and limited flow.

Looking at existing water use and allocation, the model could be used to redefine and reallocate current rights according to model optimization results for the present day. Doing this, however, would immediately affect the interests of different counties, so public acceptability would be a major issue. Hence, allocating rights based on existing patterns of water use and then allowing allocations to adjust through market trading would be an option, though this may be unrealistic given the embryonic nature of water markets in China (see below).

Defining and allocating present day rights to reflect future changes in levels and patterns of water demand, future economic development and future conservation efforts is more complex. Forecasts of future water use are generally based on integrated water use plans and would provide some scientific foundation, but forecasting is still uncertain, especially in the Chinese context where social and economic conditions are changing rapidly. Hence a process of gradual, periodic adjustment is recommended, with movement in the direction of (periodically updated) model recommendations for rights allocation.

In conclusion, action is needed to address the problems of water resources degradation in the Shiyang River Basin and many other river basins in China. In the Shiyang Basin, ecological deterioration in the lower reaches of the basin (Minqin County) is especially acute, and there is an urgent need to reallocate surface and groundwater resources in the basin. In view of the fact that water markets are not yet firmly established in China, allocating rights based on existing patterns of water use, with future adjustment towards an 'optimum' via trading, is unrealistic. Hence it is more appropriate and rational to allocate initial water rights based on detailed water resources plans and future projections of water use in the basin. Plans should be adjusted periodically until water withdrawals in the basin fall within the caps proposed by the water allocation model.

Acknowledgements

This paper is the result of a project undertaken under the auspices of the Australian Department of the Environment, Water, Heritage and the Arts and the Chinese Ministry of Water Resources, with funding provided by AusAID, the Australian Agency for International Development. The study was also supported by the "948 Project, Water Rights Reform Assessment and Key Technical Issues Study in China", funded by the Chinese Ministry of Water Resources, 2007–08.

References

(*Note: All references are in Chinese.*)

Gansu WRD and DRC (2005) *Restoration plan for the Shiyang River Basin in the near future* (Gansu: Gansu Provincial Water Resources Department and Development & Reform Commission).
Gao, E. (2006) *Water Rights System Development in China* (Beijing: China Water and Hydropower Publisher).
Ge, J. (2003) The identification of the water rights and water transfer, *China Water Resources*, 4, pp. 11–13.
Ge, M. (2002) The water rights allocation models and the water rights allocation study in the Yellow River, *ShanDong Social Science* [in Chinese], 4, pp. 35–39.
Ge, M. (2004) Study on models of water rights initial allocation, Masters Dissertation, Hohai University, China.
Liu, B. (2003) On the concept of water rights, *China Water Resources*, 1, pp. 32–33.
Lin, Y. (2002) Discussion on "initial water rights", *Zhejiang Water Science and Technology*, 5, pp. 1–10.
Wang, S. (2001) Water rights and water markets: About the economic means for the optimal allocation of the water resources, *Hydropower Energy Science*, 3, pp. 1–5.
Wang, X. (2006) Study on water allocation theory and basin evolution model in arid region, Doctoral dissertation, Tsinghua University, Beijing, China.
Wang, X., Wang, Z. & Zhao, J. (2006) Allocation model of water resources usufruct for Shiyang river basin, *Journal of Irrigation and Drainage*, 5, pp. 61–64.

A Water Rights Constitution for Hangjin Irrigation District, Inner Mongolia, China

HANG ZHENG, ZHONGJING WANG, YOU LIANG & ROGER C. CALOW

ABSTRACT   *In order to supply water for growing industrial needs in Ordos City, Hangjin Irrigation District on the south bank of the Yellow River, Inner Mongolia has traded some of its irrigation water to downstream factories. The trading is termed "irrigation water-saving supported by industrial investment, with saved water traded to industry". At the same time, Hangjin Irrigation District has conducted a comprehensive reform of irrigation water management focused on water rights. This paper describes the current status of water management in the district, outlines some of the problems water trading has produced, and presents a framework for further water rights reform focused on rights allocation, the granting of volumetrically-capped water certificates and tickets, water use planning and monitoring, and the responsibilities of water user associations in ensuring that individual farmers receive fair allocations. The paper then summarizes key recommendations of relevance to Hangjin and other irrigation districts in China.*

Background

The Inner Mongolia Autonomous Region in China enjoys exceptional advantages. In particular, the region has an abundance of natural resources for the development of mining, electric power, metallurgy, chemical and machinery processing industries. The region plans to use these resources to build a large energy base in the 'golden triangle' of Hohhot, Baotou and Ordos (Figure 1) to create an affluent society. However, the serious shortage of water resources hinders the development of the regional energy industry, and the region's allocation of water from the Yellow River Conservancy Commission (YRCC) is already fully committed. It is under such circumstances that the autonomous region initiated a pilot programme involving the transfer of water rights. Since 2003, a number of pilot projects for water rights transfer have been launched by the YRCC and the Inner Mongolia Department of Water Resources (Shen *et al.*, 2006), aimed at meeting the growing water needs of downstream industrial users.

One of the first such pilots has involved Hangjin Irrigation District. Beginning in 2004, the newly established Office of Water Rights and Transfer in Ordos City has overseen a programme in which water saved through canal lining in the district is transferred to downstream industries, with the costs of lining met directly by the industrial beneficiaries.

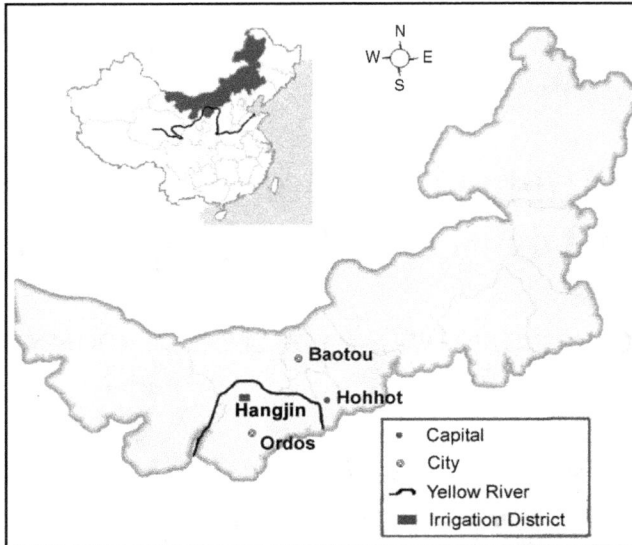

Figure 1. Map of Inner Mongolia showing Yellow River, major cities and Hangjin Irrigation District. *Source:* WET (2007).

According to the *Inner Mongolia Autonomous Region Water Rights Transfer Planning Report*, in the three-year period from 2005 to 2007, 13 enterprises invested a total of RMB 600 million in canal lining (Inner Mongolia Water Resources and Hydropower Survey & Design Institute, 2005, pp. 13–17). According to the plan, the implementation of the project will save as much as 138 million m^3 of water. Industrial users funding the capital costs of canal lining are also obliged to meet the ongoing operations and maintenance costs of canal repair over a 25-year term.

The channel lining and water transfer programme in Hangjin highlights one response to a wider problem in China—that of increasing scarcity and growing competition for water between uses and users. In this context, agriculture is under growing pressure to release water to urban and industrial users. Clear rules are needed for doing this and, increasingly, clear rights will be needed within irrigation districts (IDs) so that farmers can be confident about how much water they will get, and when they will get it. Moreover, a system of clearly defined, secure water rights provides the foundation for many other reforms aimed at managing demand and increasing efficiency, including water pricing and water trading.

Hangjin Irrigation District

Hangjin County is located to the northwest of Ordos City in Inner Mongolia (see Figure 1). Along its northern margin, the Yellow River winds down with a length of roughly 253 km, making Hangjin County the longest flowing section of Yellow River of all counties nationwide. The county includes nearly 40 000 ha of designated farmland along the Yellow River, and is one of three major irrigation zones of Inner Mongolia. It is also one of China's main grain producing areas. Hangjin Irrigation District (HID) in Hangjin County—the focus of this study—is the only irrigation district in Ordos with the right to take water from the Yellow River. HID is located on the south bank of the Yellow River and covers an area of approximately 23 000 ha. Of this, roughly 21 000 ha is gravity fed and 1700 ha is pumped (at the head of the system).

Hangjin Irrigation District draws all of its water from the Yellow River. Its water use is therefore controlled, ultimately, by the YRCC which sets minimum flow requirements for the river at provincial/regional boundaries based on an Annual Allocation Plan (Table 1), and allocates relative shares to individual provinces and regions according to supply and demand conditions. In a normal year, Inner Mongolia therefore receives 5.86 billion m^3 out of a total flow of 37 billion m^3.

Table 2 illustrates how HID—like other IDs in Inner Mongolia—operates within a volumetric permit, or cap, held by the Inner Mongolia Yellow River Irrigation Management Bureau (IM-YRIMB), under the Inner Mongolia Water Resource Department (IM-WRD). The maximum (sometimes termed 'normal') gross diversion to the district—the permitted volume—is 410 million m^3 per year, including a mandatory return flow of 35 million m^3 per year. So, the normal net diversion to HID is 375 million m^3. Return flows are fed back to the river through four main drainage channels. Savings of 130 million m^3 per year from canal lining, traded out of the irrigation district, will leave an ongoing diversion of 280 million m^3 per year, as illustrated in Figure 2.

By 30 September 2006, a total of six canal lining subprojects had been completed, each funded by a separate industrial enterprise. This has included the lining of 126 km of main canal, 13 km of sub-main main channel, 76 km of the third order canal, 69 km of fourth canal, 63 km of fifth canal and 24 km of sixth canal, with a combined investment of RMB 405 million. The outcome is a transfer of 78 million m^3 of water to downstream users (WET, 2007).

The idea of 'Industrial Investment in Water Saving for the Transfer of Agricultural Water Rights' has helped alleviate the water shortages experienced by industry, and has also helped reduce the burden of farmers by saving water and reducing farm costs. Currently, the annual water fee for each householder has been reduced by around 20–30 RMB per year (or by around 25%). Farmers' costs have reduced because they no longer have to pay for water losses in the channels that deliver water to the point where water user associations (WUAs) make bulk purchases on behalf of the farmers they represent.

The channel lining and transfer project has had many benefits. However, trading has also created a number of problems, particularly for the irrigation agency that is responsible for managing and maintaining irrigation infrastructure above WUA purchase points—the Hangjin Irrigation Management Bureau (HIMB). Moreover, the rights of farmers within the district remain ambiguous. WET (2007) summarizes the problems as follows:

- *Decline in agency income.* Trading has reduced the income of HIMB, because the agency relies on the purchase of water tickets, rather than core funding from government, to fund its activities. Specifically, as less water now needs to be delivered to farmers following canal lining, ticket sales and therefore revenue has declined.

- *Possible impacts on third parties.* The transfer may affect users outside the district by reducing leakage and hence groundwater recharge. The potential for such impacts requires further evaluation and monitoring.

- *Irrigation remains inefficient.* While the transfer of saved water may increase

Table 1. Water allocation in the Yellow River

Province/region	Qinghai	Sichuan	Gansu	Ningxia	Inner Mongolia	Shaanxi	Shanxi	Henan	Shandong	Heibei & Tianjin	Total
Annual water use billion m³	1.41	0.4	3.04	4	5.86	3.8	4.31	5.54	7	2	37
Percentage allocated to each province	3.8	0.1	8.2	10.8	15.8	10.3	11.7	15.0	18.9	5.4	100

Source: Yellow River Annual Allocation Plan for Non-flood Season, July 2006–June 2007.

Table 2. Water abstraction permits, Yellow River, Inner Mongolia, 2005

Permit no.	Abstraction license ref. no.	Diversion point	Permit holder	Gross permit ($10^4 \mathrm{m}^3$)
1	14002	South bank gate	IM - YRIMB (for HID)	41 000
2	14003	North bank (Shenwu) gate	IM - YRIMB (for Hetao ID)	45 000
3	14004	North bank (main) gate	IM - YRIMB (for Hetao ID)	432 000
4–9	14005–14010	Baogang pump station	Baotou City	57 930
10–16	14011–14018	Madihao pump station	Tuoketuo County	18 835
17	14014	Xiaoshawan pump station	Zhenger County	4 700
Total				599 465

Source: Data supplied by the Inner Mongolia Yellow River Irrigation Management Bureau.

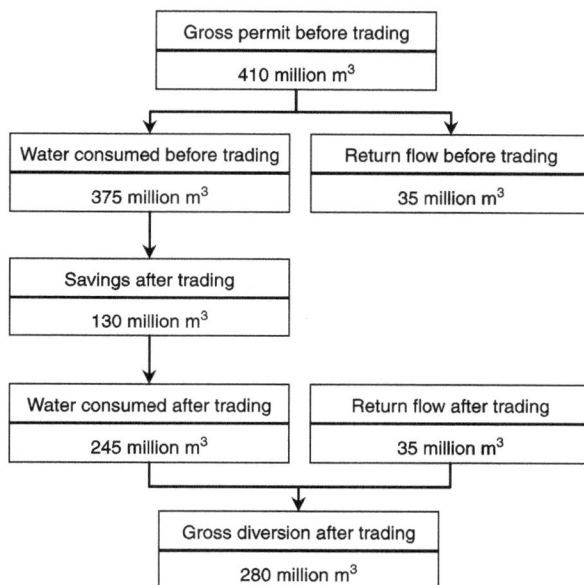

Figure 2. Diversion, consumption and return flows for Hangjin Irrigation District. *Source:* Inner Mongolia Water Resources and Hydropower Survey & Design Institute (2005).

the allocative efficiency of water between uses, it does not create incentives for more efficient use of water within the district by farmers (more 'crop per drop'). The evaporative losses from flood irrigation will continue to be significant.

- *Water rights within the district remain ambiguous.* While water diversions to HID are clearly defined through the permits issued by YRCC, the water rights of WUAs and the farmers within them are not. Specifically, farmers and WUAs continue to 'bid' for water (see below) in annual and seasonal negotiations with HIMB, rather than having defined and secure entitlement.
- *Monitoring is rudimentary.* Water monitoring is still based on equipment and methods used in the 1950s, and all recording is carried out by hand. This creates a heavy workload for agency staff, though the agency, WUA managers and farmers agree that monitoring is transparent and consistent.

This paper focuses principally on the final two issues, specifically water rights definition, allocation and monitoring within HID.

A Framework for Irrigation District Water Rights

A framework for a modern system of volumetrically defined water rights in HID has been developed (WET, 2007). It is proposed that this serve as a template for guiding reform in other IDs in China, as competition for water increases and agricultural users face growing pressure to account for their water and release 'surpluses' to urban and industrial users. The sections below discuss rights definition, allocation and management issues within HID. The principal focus is on improving the distribution of water within an ID so that farmers receive secure, transparent and equitable allocations within the overall permitted allowance of the ID.

Initial Water Rights Allocation

Drawing on fieldwork conducted in HID, WET (2007) describes how the water diverted to the district under its irrigation permit is currently allocated through main and branch canals, and down to individual farm households. In common with many IDs in water-scare northern China, the allocation process combines bulk volumetric charging to farmer groups (increasingly WUAs) established on branch canals, with area-based charging for farmers. Water User Associations purchase pre-paid water tickets on behalf of farmers, and are responsible for (amongst other things) distributing water within their command areas and collecting fees.

WET (2007) describes how water allocation to WUAs could be improved according to the principles of fairness, efficiency and environmental sustainability, amplified below. They also describe how the water rights of WUAs could be volumetrically defined and capped through the issue of Group Water Entitlements (GWEs) at the point at which WUAs pay for bulk deliveries. Below this point, farmers would continue to pay for water on an area basis, as delivery and monitoring infrastructure in Hangjin, and most IDs in China, is not in place to monitor individual entitlements at the household level.

A volumetric cap on the water rights of WUAs needs to fully consider existing patterns of water use within and between WUAs, and the experience of farmers, WUA representatives and HIMB staff in administering present systems. Hence it is proposed that rights allocation follows existing practice by linking land and water rights. In other words, rights assigned would be directly linked to the (existing) irrigated areas of each WUA, and could not be negotiated upwards by a WUA seeking to expand its irrigated area or plant more water-intensive crops, for example. Thus, one objective of defining and enforcing WUA-based GWEs would be to end the requirements approach to water use planning that currently prevails so that, in future, water savings rather than additional supply would be used to maintain or increase farm production and farmer incomes.

Different regions and different groups of people should enjoy equal rights to water for survival and development. Hence the allocation of rights should guarantee fairness between different management sections of an ID, different WUAs and different water users and, in particular, afford protection to those farmers with small land holdings. In defining and allocating rights, consideration should also be given to 'third party' impacts on (linked) environmental services and other downstream users, such as groundwater users dependent on return flows from the irrigation district.

How can the GWEs of individual WUAs be calculated to account for these factors, and to account for channel losses incurred to the points in the system at which WUAs purchase water? WET (2007) describes the calculations involved. The combined irrigated area of all 43 WUAs in the gravity flow section of HID is estimated at 21 322 ha. The total volume of water that needs to be delivered to fourth level sluices (and therefore WUAs), after subtracting losses in the canals above, is estimated at 143 million m^3 per year. The total volume of water that needs to be diverted from the Yellow River to meet WUA requirements and cover conveyance losses is 225 million m^3 per year. Total losses in the canals above fourth level sluices are estimated at 82 million m^3 per year. Using these data, and similar calculations covering allocations to individual WUAs, the long-term initial water rights of each WUA in HID can be determined as GWEs. These, in turn, form the basis for the issue of water certificates (see below).

In contrast to the current farmer-driven approach to estimating water needs in ID, such an allocation provides a more scientifically sound basis for defining and capping rights within the overall allowance of the irrigation district, and for accounting for all transmission losses through main and branch canals to WUAs. Since losses in each canal have now been estimated, future conservation efforts—including trading in transmission savings—can be better targeted and quantified. In this way, the approach to defining and allocating GWEs described above can form the basis for rights reform in other IDs.

Water Rights Certificates and Water Tickets

A system of water rights certificates can be used to formalize the rights of WUAs, providing information on long-term rights (defined by GWEs), annual water entitlements (defined by available supply in any given year) and the water purchased in each irrigation period. In addition, the system can provide information on any water transactions that have occurred between WUAs, and between WUAs and the irrigation management agency. Table 3 provides a summary of certificate functions and uses.

To establish and operate such a system, the following steps are proposed (WET, 2007):

- After an initial water rights allocation process, the irrigation agency grants rights to each WUA in the form of a water certificate. This will show each WUA's long-term water right.
- At the beginning of each year, the agency calculates the proportional water share that each WUA is entitled to (an annual entitlement) based on expected water availability in that year.
- Before each irrigation period, the agency adjusts, as necessary, each WUA's annual entitlement in light of predicted supply to give a corresponding water purchase limit for all remaining irrigation periods. The purchase limit is recorded on each WUA's water certificate.
- After purchasing water tickets in any given irrigation period, the purchase amount is recorded on the certificate to calculate the remaining purchase allowance, or entitlement, of the WUA for the next period. In other words, a process of continuous water accounting is adopted between irrigation periods.

Table 3. Functions and uses of water certificates

Function	Use
Voucher for long-term rights	The irrigation management agency records each WUA's long-term water rights (Group Water Entitlements) in a water certificate.
Calculation of purchase limits	At the beginning of the year, the irrigation management agency calculates the water purchase limit (annual entitlement) of each WUA and records this information on the certificate. After purchasing tickets in each irrigation period, the purchase amount will be recorded on the certificate to calculate the remaining purchase limit for the following periods. WUAs can purchase tickets up to the limit.
Record of water trading	The irrigation management agency records all information on water transactions.
Reference for water rights reallocation	The irrigation management agency will accumulate data on actual water use across seasons and between years, helping to guide any future adjustment.

- Any water trading is recorded by the relevant agency section office on the water certificates of both buyer and seller. Trading with other sections is also checked and registered with the agency. Certificates would also show actual water deliveries after trading.

After a reasonable period of operation (5–10 years), the irrigation management agency can review certificates in light of actual water use and trading experience, and revise as necessary. Following any long-term trade of water rights, the irrigation management agency can take back old certificates and issue new ones after thorough auditing and recording.

For each WUA's purchase of water, it is proposed that the current system of pre-payment through water tickets is continued. Water tickets provide the basis for water purchase, water delivery and water trading within prescribed limits. The ticketing system can ensure that both WUAs and the irrigation management agency have clear information on prices, deliveries and volumetric rights, allowing WUAs to trade savings freely (Wu & Wu, 1993). Water User Associations would buy water tickets according to their water certificates before each irrigation, and would also be allowed to purchase extra water from those WUAs deciding not to use their full allowance (Feng & Li, 2006). Table 4 provides a summary of ticket functions and uses.

In summary, water rights certificates would formalize the long-term water rights of WUAs within an ID. Water tickets would then 'translate' these rights into real-time rights for WUAs, allowing them to purchase water within the cap for a specific period, and according to how much water has been purchased previously. Long-term and real-time water rights are then connected through water use planning, which converts long-term GWE into the real-time water cap and water use scheduling according to the planned water demand and the runoff forecast of the river. The relationship between water rights, water rights certificates and water tickets is shown in Figure 3.

Table 4. Functions and uses of water tickets

Function	Use
Support for permit control and quota management	WUAs buy tickets up to their caps; Hangjin Irrigation Management Bureau (HIMB) sells tickets according to water availability and water rights limits.
Pre-payment for water	Water is only supplied by HIMB once WUAs have purchased tickets.
Water trading and monitoring	WUAs can buy and sell 'saved' tickets; HIMB monitors ticket turnover and adjusts caps as necessary.
Payment voucher – rights and duties	Tickets provide information on Group Water Entitlements, actual delivery and payment—a summary of entitlement and payment obligation.

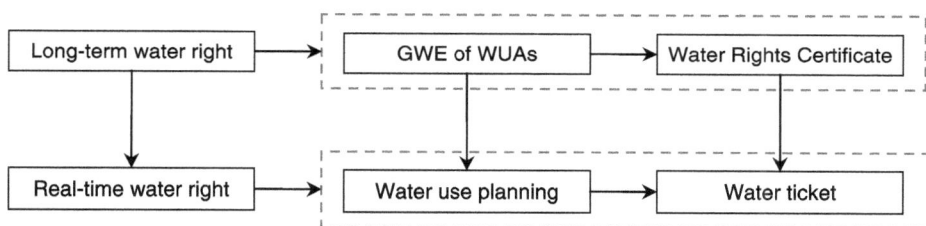

Figure 3. Relationship between water rights certificates and water tickets.

Water Use Planning

The objective of water use planning is to schedule water diversion, storage, delivery and use in an ID according to the requirements of farmers, available supply from the river, and flow through the irrigation channel system. A water use plan is a guideline for the rational delivery and use of water within an ID, and can help improve irrigation efficiency and save water. In this paper, it is proposed that the water use plan takes the GWEs discussed previously as a starting point, and then translates them into a real-time irrigation schedule for WUAs. WET (2007) proposes that this occurs through a computer-based model that can balance demand and supply, guide allocation between WUAs and help manage rights in a quick and transparent manner.

At the beginning of the year, the annual water use plan for the ID would be prepared by the irrigation management agency, based on the annual water use plans submitted by each WUA (within capped limits), and submitted upwards through the irrigation agency to the higher-level department for approval, such as the river basin management department. The river basin management department would then revise and approve the annual available water cap and the water scheduling of the ID, according to the water abstraction permit of the ID and the annual runoff forecast of the river. Afterwards, the irrigation district management agency would adjust the annual plan accordingly, and announce it to WUAs.

Prior to each irrigation, a WUA would then prepare and submit a plan for that period to the irrigation management agency for approval. The agency would check the available water allowance for each WUA, accounting for previous purchases, and make any necessary revisions or suggestions, under the cap and overall irrigation scheduling. Following ticket purchase, a final water use plan would be confirmed in accordance with sold ticket volumes and the scheduling needs of all WUAs. The computer model would help managers prepare, modify, summarize and publish schedules, and could be interrogated quickly by all relevant stakeholders. The model would also help managers deal with the effects of runoff variation and hydrological uncertainty, including emergency planning in the event of floods or droughts.

Water Users Associations

A key element of irrigation reform is the promotion of WUAs as farmer-run, participatory institutions that take the place of village leader-run water control organizations or government agencies, and take over management of water allocation and infrastructure management at a local level (Wang *et al.*, 2006). Water User Associations are registered as legal entities under Chinese Company Law.

In HID, a total of 43 WUAs have been established since 2000 under third level canals in the gravity flow sections, with a further 40 planned for completion by the end of 2008. The boundaries of WUAs are defined by areas irrigated by tertiary and fourth canals. As a result, WUA and village boundaries do not always match. HIMB works with WUAs on the development of Annual Water Allocation Plans and scheduling arrangements, and WUAs are obliged to purchase water tickets prior to each irrigation period. It is proposed that WUAs hold and democratically manage GWEs on behalf of farmers and, within capped limits, continue to develop scheduling plans for household members, collect water fees, purchase water tickets from the ID management agency and undertake maintenance work on the infrastructure within their command areas.

The ability of WUAs in Hangjin (and elsewhere in China) to manage water rights effectively under capped GWEs depends on a number of different factors. WET (2007) identifies four key pre-conditions, based on a survey of WUAs and farm households conducted in 2007. First, GWE-based accounting through water certificates would need to be carefully monitored and enforced. The allocation system in HID combines bulk volumetric charging to WUAs established on branch canals, with area-based charging for farmers. Under such a system, the irrigation district management agency supplies water to WUAs on a contractual basis; contracts have no (current) legal authorization, but do specify the rights and obligations of both the agency and WUAs. Such contracts, or agreements, provide a type of group water right, albeit one of limited security. In Hangjin, moreover, the delivery of water to WUAs is governed by service contracts between WUAs and HIMB. Fieldwork in HID (WET, 2007) suggests that these arrangements provide a sound basis for clarifying rights and responsibilities around water delivery and payment, and for the monitoring and recording of delivery and payment. They are recommended for other irrigation districts embarking on quota-based rights reform.

Secondly, infrastructure needs to be compatible with defined rights and local management capacity. Any discussion on water rights reform cannot be isolated from an understanding of the infrastructure that is available to deliver, monitor and record water flows. In Hangjin, and in most other IDs in China, irrigation systems have not been designed to deliver and record flows to individual farmers. In these circumstances, volumetric rights can only be defined, monitored and enforced down to the level of the WUA and, conceivably, to production teams managing tertiary canals. Hence in such systems it is proposed that capped rights are allocated to WUAs through GWE-based certificates, recognizing that farmer-level entitlements cannot (yet) be implemented.

Thirdly, WUAs need well-specified management functions, authority and account-ability. A key issue here is whether WUAs genuinely represent the interests of all farmers, and whether they have the capacity to resolve competing claims and disputes. In Hangjin and other IDs where WUAs have been established, the management functions and authority of WUAs are spelt out in a charter, or a set of written rules. The ability of farmers to assert individual claims within the bulk GWE will therefore depend on whether WUAs act as genuine organs of democratic self-management, and whether elections required under their charter are held in an open, inclusive and fair way. It is therefore suggested that the democratic management of WUAs is scrutinized closely by the ID management agency for a period of time after initial establishment. Periodic audits of WUA performance covering this and other tasks (e.g. financial book-keeping) are recommended.

Finally, WUAs require adequate resources. A common assumption in irrigation turnover programmes is that WUAs are better (than government agencies) at undertaking water allocation, distribution and fee collection in a cost-effective way. However, new obligations may be a serious burden on WUAs if they have been formed without adequate attention to their ongoing support needs. A key question in Hangjin and other IDs, therefore, is whether pressure to reduce government outlays—a key factor driving management transfer—has extended to an unwillingness to provide sufficient resources for WUAs to retain elected staff and carry out management tasks effectively, particularly in relation to long-term water allocation, technical backstopping and maintenance. It is therefore recommended that WUAs are allowed to retain enough ticket revenue to cover the salary costs of their full-time staff, and to cover operation and maintenance tasks within the WUA command area. Resourcing issues could be similarly monitored through periodic audit.

Water Metering and Monitoring Systems

Many existing monitoring systems in China are crude, and need to be upgraded to support the operation and management of a modern water rights system. In HID, for example, water levels are measured using simple gauges, and flows are measured with traditional flow meters. All measurements are done by hand, with staff having to monitor and regulate flows through over 20 gates to WUAs. In a large ID this creates a very heavy workload for staff and at times of peak water demand, there may be a shortage of manpower.

Future pressure on IDs to release water for urban and industrial users may increase pressure for more accurate monitoring of allocations to WUAs. In this context, automated water monitoring systems may help solve current and future problems, saving labour and money and providing more accurate monitoring and regulation of increasingly scare water. The design and use criteria that a monitoring system needs to meet are outlined below (WET, 2007).

- *Automated monitoring and data transmission.* Automated systems are more accurate and less labour intensive than manual ones, eliminating the need for station staff to travel between and monitor individual sites.
- *Rapid calculation and easy access to data.* Data calculation and analysis should be quick and accurate, and data interrogation should be simple and direct. At present, data enquiries in HID can only be answered by sifting through large numbers of paper records.
- *Remote control and monitoring of main sluices.* The irrigation management bureau should be able to operate sluices on the main and branch canals at least remotely, avoiding long distance travel for station staff and the need to spend many hours at individual sites.
- *Transparency.* It is important that an automated system retains the transparency of the existing system. In particular, WUA managers and farmers should have easy access to information on water deliveries to WUAs to build confidence in the quota-based certificate and ticketing arrangements.
- *Affordability.* Any upgraded system needs to be affordable in terms of both capital costs and the ongoing costs of repair and maintenance. Benefits can help offset costs, however, and are likely to include time (labour) savings for irrigation management agency, and water security-income gains for farmers (through more timely and reliable water delivery).
- *Durability and security.* An upgraded system must be able to cope with the sediment-laden inflows of the river, and not require constant adjustment and maintenance. It should also be equipped with alarms to increase security, and data security and virus protection should be included.
- *Ease of use.* Advanced systems must be capable of being operated and maintained by station staff.

Towards an Integrated Framework for Rights Management in Irrigation Districts

Drawing on the discussion above, a broad water rights framework is proposed for HID and other IDs in China. The framework consists of three elements: institutions, irrigation services and regulations. These are described briefly below and illustrated in Figure 4.

The institutional component refers to the management institutions responsible for water allocation and delivery, including the relevant river basin management departments, ID management agencies and WUAs. The government river basin management department is responsible for allocating water and issuing water permits to IDs, and auditing their water use plans. No changes to existing allocation arrangements and responsibilities are proposed here. Irrigation management agencies are mainly responsible for water allocation to WUAs. In this paper, it is proposed that they assume responsibility for the granting and overall management of water rights certificates and water tickets issued to WUAs, in addition to existing responsibilities for collecting water fees, preparing the water use plan of the irrigation district, and monitoring water deliveries to WUAs. Water User Associations, in turn, would assume responsibility for purchasing water tickets within the caps set by GWE calculations, and would manage and monitor allocations under the cap to individual farmers. Field investigations in Hangjin suggest that, where ticket-based payment and contracting systems are already established, the capped

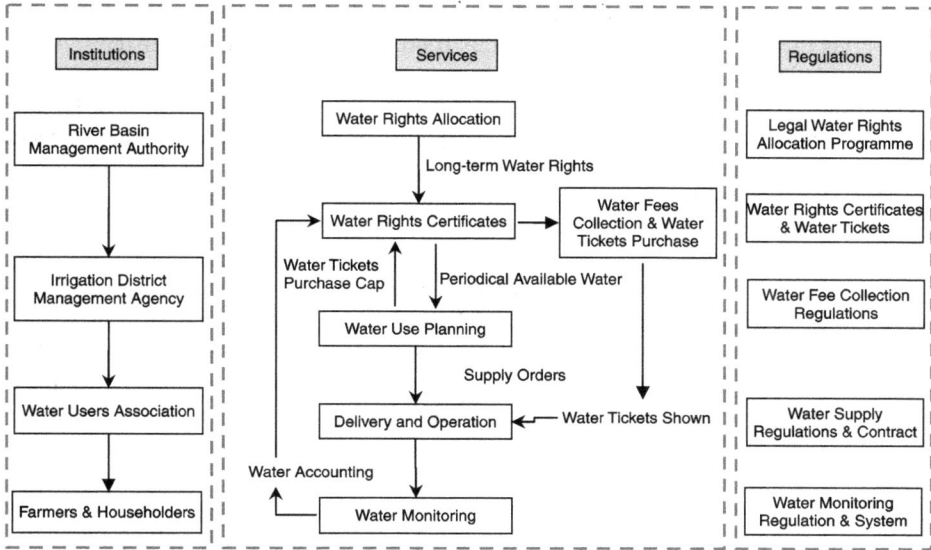

Figure 4. A framework for water rights in an irrigation district.

arrangements for allocating and purchasing water proposed in this paper could be implemented fairly easily.

Irrigation services include the initial allocation of water rights, the issue of water certificates and tickets, water use planning, water delivery and operation of infrastructure. The permitted water abstraction volume of the whole irrigation district is allocated to WUAs through the initial water rights allocation process described, forming the basis for granting water rights certificates and the sale of water tickets. WUAs would purchase tickets within their allocated rights, prepare a water use plan and submit it to the irrigation district management agency for approval. The irrigation district management agency would then complete a water use plan for the whole district and issue delivery instructions to sluice operators, according to each WUA's water use plan and remaining ticket purchase allowance. Deliveries would be monitored and signed off as they are now, with agency staff and WUA managers entering into seasonal contracts, and jointly monitoring and confirming allocations. The irrigation district management agency would record each WUA's available water, purchased water and supplied water every year, and every watering in their water rights certificates on a continual basis, in order to check the water account and guide water supply in the next period. Regulations would then ensure effective implementation and monitoring of the services above, and would need to cover management regulations for the issue and use of water rights certificates and water tickets, water fee collection, water delivery and water monitoring. All management regulations and systems need to be carefully coordinated.

Conclusions and Recommendations

Based on field investigations in HID, a water rights framework for IDs in China has been proposed in this paper, based on an initial water rights allocation, the issue of water rights certificates, sale of water tickets, water use planning and effective management of farmer-level rights through WUAs. Drawing on this framework, the authors offer the following recommendations:

(1) Group Water Entitlements should be defined and allocated to WUAs in HID and other IDs, and could additionally be given legal basis by the government so that rights can be legally asserted and defended, providing greater security to WUAs and farmers. In addition, a water rights management system should be developed for all IDs, including regulations that cover water use planning, water delivery, emergency planning and risk management, the collection of water fees and maintenance of infrastructure. Entitlement-based allocation planning underpins future water conservation efforts and the development of a modern, socialist countryside in China.

(2) The use of an allocation plan to allocate water to WUAs in HID and other IDs is feasible. The annual allocation process in an ID needs to define and allocate GWEs within the overall permitted allowance of the district, determined by the relevant river basin authority. Allocation planning of this kind is more fair and transparent than existing arrangements.

(3) Existing contract and ticketing procedures operating between HIMB and WUAs are well understood and respected. They provide an excellent platform for the introduction of GWEs and ticket-linked water certificates. Those WUAs that have set up systems of continuous water accounting between irrigations, and volumetric delivery to (and billing of) individual production teams, will be better able to meet new quota obligations in a fair and transparent manner. Such systems are recommended for other IDs in China embarking on rights-based reform.

(4) Water trading to downstream industrial users has reduced the revenue available to HIMB. The issue of funding will need to be addressed to ensure the long-term sustainability of the trading programme and channel infrastructure, and to protect farmers' long-term water rights. Management and institutional reforms in the ID should be conducted as soon as possible to improve management of the channels, enhance the financial position of the irrigation agency and secure new investment and financial resources. Most importantly, funding for the maintenance of newly lined channels in Hangjin should be secured from industrial enterprises as soon as possible. Similar channel lining and water transfer initiatives being considered by government agencies for other IDs in China need to learn from the experience of Hangjin.

(5) Information and monitoring systems in Hangjin and other IDs need to be gradually upgraded to improve accuracy and reliability and reduce manpower requirements. A key priority is to strengthen monitoring of water deliveries at WUA purchase points, as monitoring here affects both WUA payment and compliance with any new system of GWE-based water rights certificates.

Acknowledgements

The authors would like to thank Mr Xiaoyou Sheng, from the Inner Mongolia Water Resources Department, and Mr Ruichun Liu and Mr Baijian Ping from the Hangjin Irrigation Management Bureau, who supported the investigations. This paper is the result of a project undertaken under the auspices of the Australian Department of the Environment, Water, Heritage and the Arts and the Chinese Ministry of Water Resources, with funding provided by AusAID, the Australian Agency for International Development. The study was also supported by the "948 Project, Water Rights Reform Assessment and Key Technical Issues Study in China", funded by the Chinese Ministry of Water Resources, 2007–08.

References

Decree on Yellow River Water Resources Regulation (2005) Yellow River Conservancy Commission. Available at: http://www.yellowriver.gov.cn/ziliao/zcfg/fagui/200612/t20061222_8784.htm (accessed 21 December 2008).

Feng, E. & Li, X. (2006) The study on the water rights system in the irrigation districts in the inland river basin, China, *Gansu Agricultural University Journal* [in Chinese] 2, pp. 105–108.

Inner Mongolia Water Resources and Hydropower Survey & Design Institute (2005) *Inner Mongolia Autonomous Region Water Rights Transfer Planning Report [in Chinese]*, Technical Report.

Shen, D., Sheng, X., Wang, R. & Liu, A. (2006) Water rights reform in Inner Mongolia, China, *China Water Resource* [in Chinese], 21, pp. 9–11.

Wu, D. & Wu, Z. (1993) On water use planning and water tickets, *Water Economy* [in Chinese], 8, pp. 50–52.

Wang, X., Huang, L. & Rozelle, S. (2006) Incentives to managers or participation of farmers in China's irrigation systems: which matters most for water savings, farmer income and poverty? *Agricultural Economics*, 34, pp. 315–330.

WET (2007) *Water Entitlements and Trading Project (WET Phase 2) Final Report* December 2007 [in English and Chinese] (Beijing: Ministry of Water Resources, People's Republic of China and Canberra: Department of the Environment, Water, Heritage and the Arts, Australian Government). Available at: http://www.environment.gov.au/water/action/international/wet2.html.

A Comparison of Water Rights Systems in China and Australia

ROBERT SPEED

ABSTRACT *This paper describes and compares the reforms in China and Australia associated with granting water users better defined, more secure and (often) tradable entitlements to water. The paper considers the lessons that each country may learn from, and teach to, the other. The paper discusses policy issues and solutions in both countries in respect of: risk sharing and compensation for changes to rights; environmental flows; trans-jurisdictional approaches to water rights; trading water rights; and integrating water rights within the broader water supply and catchment management framework.*

Introduction

Australia has been undertaking major reforms to its water sector for more than 15 years. These reforms have involved institutional restructuring, the recognition of environmental requirements for water and the adoption of 'water rights'—with water users granted better defined, more secure and (often) tradable entitlements to water. China is also in the process of implementing major changes to the way it manages water, including the adoption of a 'water rights transfer system' (State Council, 2006). Many of the management issues that China is currently seeking to address are problems faced in Australia over the past decade. Meanwhile for the first time in Australia, the federal government is assuming primary responsibility from the states for water management in Australia's largest river basin, the Murray-Darling; such an approach is not new in China, where the central government has a long history of controlling trans-provincial waterways. As such, there are significant lessons the countries can learn from one another.

This paper compares and contrasts the evolution of water rights systems in China and Australia. It considers the challenges being faced in each country and the policy options being implemented to address common problems. It considers how differences in the political and institutional arrangements have led to different approaches and identifies lessons from both countries in implementing water rights systems. The paper starts with an overview of water rights systems and water rights theory. It then compares and contrasts the political, social and environmental contexts, as well as the management systems in place, in Australia and China. The paper then describes the water reform processes that each country has undertaken over the past decade or more. The paper concludes with a discussion of some of the key issues to have emerged.

It is important to note that this paper is not intended to assess the water management performances of the two countries against one another. The differences in levels of development and, particularly, the challenges China faces in providing food and water for the world's largest population mean that such an exercise would be of little practical value. Instead, the aim is to highlight some of the common challenges as well as the different policy options being adopted.

The Theory of Water Rights

A water rights system is, fundamentally, one that identifies the total available resource and divides it, via the grant of water rights, amongst different users (Xie, 2008). A number of studies have identified the key requirements of a water rights system (Productivity Commission, 2003). These generally include:

- universality: all water resources are covered by the system of rights;
- predictability of volume: users have a reasonable expectation of the volume of water that will be available to them;
- enforceability: the right is protected from encroachment by others;
- certainty of title: there is legal recognition and protection of rights;
- duration: the time period over which users possess the right is specified;
- exclusivity: the benefits and costs of the right accrue to the owner;
- detached from land title and use restrictions: the right is separate and free of any requirements to hold land or any restrictions on how the right may be exercised; and
- divisibility and transferability: the right may be subdivided and is freely tradeable to others (Productivity Commission, 2003).

By clearly defining both the total resource and the shares of different users, a water rights system does two things. In terms of the public interest, it can provide confidence that the level of water abstractions is sustainable. In terms of private interests, it provides some certainty to water users as to what water will be available to them, both over the long term and in any given year. This in turn can provide users with confidence to plan for the future and to invest in water-dependent activities. It also reduces the potential for conflicts, as sharing arrangements are transparent and settled in advance, and provides incentives for more efficient use of resources (Productivity Commission, 2003; Xie, 2008).

A water rights system, by defining the shares of different parties, can also support a number of water management policy tools, including the following:

- Government-facilitated reallocations: where government wishes to reallocate water to different users, sectors or regions, a water rights system allows this to happen in a more transparent and (potentially) equitable manner.
- Reducing total water abstractions: likewise, if it is necessary to reduce the total volume of water allocated for consumptive purposes, this can be done in a more transparent and equitable way if the rights of water users are already clearly defined.
- Pricing mechanisms: it is simpler to apply pricing tools to water users (e.g. to encourage water use efficiency) where what they are paying for is well defined.

● Water markets and trading of water rights: water rights are a requirement for a market-based water trading system, as parties need to know what they are buying and selling, and to have confidence that the rights they buy will be protected (Xie, 2008).

It is important to note that in this article the term 'water rights' is used broadly. The term applies to a range of interests in water, including both long-term rights granted to water users under a licence or other instrument (referred to here as 'water access entitlements' or 'entitlements') and the volume of water allocated periodically (e.g. monthly or annually) to the holder of a water access entitlement, referred to in this article as 'water allocations'.

The Nature of Water Rights

Much of the academic discussion on water rights tends to focus on their legal status, and distinctions between ownership and use rights and between private and public rights. More important than the characterization of the right is what the legal and institutional arrangements surrounding and defining the right mean for the right holder, other water users and the government (as resource regulator). This includes the rules that determine how much water the right holder will receive under different circumstances and under what circumstances the right can be varied (Xie, 2008)

Importantly, water rights systems typically grant an interest in water that is more clearly defined and better protected than the interests granted under alternate management arrangements. There is arguably, however, no threshold at which an interest or entitlement to water should be viewed as a 'right'.[1] As such, the development of a 'water rights system' should be seen as a progression along a continuum: moving from one end—where the interests of different parties in water resources are poorly defined and subject to change with little or no notice or compensation—towards the other end, where interests are clearer and better protected from adverse impacts (Figure 1). The discussion below about the development of water rights in China and Australia is undertaken in this light: it describes the reasons, the ways, and the success (or otherwise) of each of the countries as they have moved along this continuum.

China and Australia Compared and Contrasted

China and Australia are physically large countries, the fourth (China) and sixth (Australia) largest in the world. Both have a diversity of river and climatic systems, including tropical,

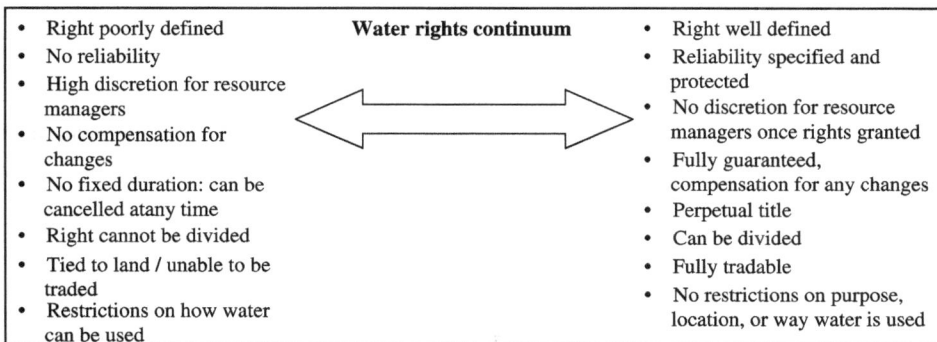

• Right poorly defined	**Water rights continuum**	• Right well defined
• No reliability		• Reliability specified and protected
• High discretion for resource managers		• No discretion for resource managers once rights granted
• No compensation for changes		• Fully guaranteed, compensation for any changes
• No fixed duration: can be cancelled at any time		• Perpetual title
• Right cannot be divided		• Can be divided
• Tied to land / unable to be traded		• Fully tradable
• Restrictions on how water can be used		• No restrictions on purpose, location, or way water is used

Figure 1. Water rights continuum.

subtropical, semi-arid, arid, inland and coastal systems. There are regions in both that suffer major water shortages, as well as parts with regular monsoons and accompanying floods (SBS, 2006).

In political terms there are stark differences. Australia is a liberal democracy and exists as a commonwealth of federated states, with the Australian Constitution passing only limited powers to the federal government. This has meant that where coordination between the federal and state governments is required, this must be done via negotiated agreements, rather than a directive from above. In practice, agreements have often included significant financial incentives from the federal government, which controls the majority of tax revenue.

In contrast, China is a socialist republic, where the Communist Party leads a single-party state, and provincial governments are subordinate to the central government. As such, the central government is able to direct the laws, policies and activities of provincial governments. Regardless of this, the nature of politics means that some level of negotiation and agreement is usually involved in any central–provincial interactions. Certainly, in both China and Australia there are tensions between the central and regional levels of government, over funding, government priorities and the local application of policies and laws.

China's population of 1.3 billion people is around 60 times that of Australia (SBS, 2006). While both countries are major primary producers, their agricultural (and irrigation) sectors are drastically different. Around 50% of China's immense population is engaged in agriculture, compared with less than 5% in Australia (SBS, 2006). As a result, a medium-sized irrigation district in China can be home to thousands of farmers, while the same area (and volume of water) in Australia might support a farm owned and operated by a single family.

As the driest continent, Australia's mean annual rainfall of approximately 400 mm is significantly less than China's 600 mm. This difference is reflected in other national water statistics (MWR, 2007; SBS, 2006). That said, there can be significant parallels between

Table 1. Comparison of Murray-Darling Basin and Yellow River Basin

	Murray-Darling Basin	Yellow River Basin
Area	1 059 000 km^2	795 000 km^2
Length	Darling (2740 km), Murray (2530 km) and Murrumbidgee (1690 km)	5464 km
Annual precipitation	530 billion m^3	370 billion m^3
Annual mean runoff	21 billion m^3	58 billion m^3
Water consumption	9.3 billion m^3 (incl. 1 billion m^3 of groundwater)	30.7 billion m^3 (incl. 9.7 billion m^3 of groundwater)
Irrigated area	1.65 million hectares	7.3 million hectares
Gross agricultural value	AU$15 billion	AU$16 billion[a]
Population	2 million (10% of national population)	172 million[b] (13% of national population)

Notes: a. RMB103 billion (converted to AUD as at July 2007). b. This includes those living in the area affected by flooding from the Yellow River. Population in the basin proper is approximately 98 million. *Source:* Yellow River Commission (n.d.); ABS (2007).

the climatic and hydrological conditions at a basin level. The comparison in Table 1 of the Murray-Darling Basin and the Yellow River Basin highlights both the similarities and differences between the countries. Due to a lack of suitable sites and fewer water supply pressures, Australia's 83 billion m^3 storage capacity is one-eighth that of China's (MWR, 2007; NWC, 2005). In contrast though, Australia has an estimated 2 million off-stream 'farm dams', constructed and operated by individual farmers, which are less common in China (NWC, 2005).

Finally, the key water-related issues in each country differ. In China, water scarcity, water pollution, flooding and soil erosion are the key water management challenges (Sun, 2002). Water pollution issues relate particularly to point source pollution from industry and a low wastewater treatment rate (OECD, 2007). In Australia, water scarcity is the greatest challenge, with salinity (especially in the Murray-Darling system) and non-point source pollution also key issues. Australia's hydrology, topography and population density mean that it does not have China's major flood management issues, or the major opportunities for hydropower development. Consequently, in Australia, most, but by no means all, reservoirs are operated with a primary focus on water supply, while China's reservoirs have a number of competing functions, including water supply, flood control and hydropower generation. The water resources management arrangements in both countries reflect, not surprisingly, the history, hydrology and political setting of each.

Water Reform in Australia

Under Australia's constitutional arrangements, the state governments have primary responsibility for water resources development and management. Historically, this has consisted of legislation vesting ownership of water resources in the state, licensing systems to regulate access to water and management via government agencies originally with an engineering and construction focus. Dams and irrigation schemes were built by government, with the cost of water supply schemes heavily subsidized (Parker, 2006).

A major turning point for water management came in 1991, when a toxic algal bloom over 1000 km long extended across the Murray-Darling Basin. The incident highlighted the deteriorating state of the basin and gave momentum to efforts to improve its management and overall health. At the same time, rapid development of water resources coupled with extended drought through the early 1990s highlighted shortcomings in existing management arrangements (Parker, 2006).

Up until the 1980s/1990s (depending on the state) water entitlements were attached to land and could not be transferred. As such, in fully allocated systems there was limited capacity for new enterprises to access water. At the same time, existing entitlement holders had few incentives to use their allocation efficiently. Water entitlements were often defined with reference to infrastructure (e.g. a certain size pump) rather than volume and reliability, and entitlements could be amended or cancelled without compensation. An incremental approach to licensing meant applications were considered on an individual basis without reference to the overall availability of water in the system and consequently the reliability of existing entitlement holders was being steadily eroded.

Annual rules applied by government agencies for sharing water were often not documented and institutional arrangements were often mixed, with the same agency at times holding the conflicting responsibilities of resource manager, water service provider and regulator.

The cost of managing water infrastructure was not economically sustainable, and environmental requirements for water were not recognized (Parker, 2006; Banyard & Kwaymullina, 2000; COAG, 1993).

Together, these issues were resulting in growing conflicts over water and threatening investment in water-dependent industries. In 1994, the federal, state and territory governments, under the auspices of the Council of Australian Governments (COAG), agreed to a water reform agenda to address these issues (NCC, 1994). Under the agreement, the states committed to a series of reforms to the water sector. These included implementing a comprehensive system of water entitlements backed by separation of water property rights from land title, allowing for trading of water entitlements, and making allocations of water for the environment as a legitimate user of water, based on the best scientific information available (NCC, 1994). Notably, the states were given financial incentives, totalling hundreds of millions of dollars, to fulfil their commitments. More recently, in 2004, the same parties entered the National Water Initiative (NWI), which reaffirmed and built upon the reform agenda set out in the 1994 COAG Agreement (COAG, 2004).

While individual states have taken different paths in achieving these objectives, implementation of the reforms has principally been through:

- Adopting catchment-based statutory 'water resources plans', to define the sustainable limits of a catchment, cap total water abstractions and provide water for the environment;
- Granting volumetric 'water access entitlements' to individual abstractors, in accordance with the limits set by the catchment plan, with defined levels of reliability;
- Specifying the rules for determining annual 'water allocations' for the holders of water access entitlements, based on annual availability;
- Allowing for trading of water access entitlements and water allocations between users, in accordance with rules to protect other users and the environment from adverse impacts; and
- Providing for compensation where changes are made that affect the value of an entitlement. (See, for example, Water Act 1989 (Victoria); Water Act 2000 (Queensland); Water Management Act 2000 (New South Wales.)

Together, these elements represent the shift to a rights-based approach to water resources management in Australia.

In recent years, the Murray-Darling Basin has experienced the worst drought since rainfall and runoff records have been collected. Many water access entitlement holders have received no or substantially reduced annual water allocations, and there are great concerns for the ecological health of the basin (CSIRO, 2008). At the same time, urban water supplies have been at record lows for many of the state capitals, including Adelaide in South Australia which takes substantial volumes of its requirement from towards the downstream end of the Murray River. In response the Australian government has developed a strategy to secure Australia's long-term water supply, which includes an investment of AU$12.9 billion over 10 years in water buybacks, infrastructure and policy reforms, including AU$3.1 billion to purchase water entitlements in the Murray-Darling Basin to be returned to the river for environmental purposes (DEWHA, 2008). At the same time, an agreement was reached and the states referred the federal government specific power over water resources in the Murray-Darling Basin. This referral provided the constitutional basis for an amendment to the

Water Act 2007 (Commonwealth), the new Federal water law. The Act establishes the Murray-Darling Basin Authority (which replaces the former basin commission) which is empowered to prepare, for the first time, a whole-of-basin water allocation plan, to be completed by 2011. The states remain responsible for catchment-level water resource plans, which will be required to be consistent with the basin plan.

Water Reform in China

China has a long history of water resources development and management. Water diversions for irrigation date as far back as 316 BC (Wouters *et al.*, 2004), and the establishment of trans-provincial agencies to coordinate flood control and protection measures go back to the 13th Century (Shen, 2004). Historically, the state (in its various forms) has played the lead role in water resources development, including a major expansion of reservoirs and irrigation districts during the 1950s and 1960s under the direction of the government of the (then) newly formed People's Republic (Calow *et al.*, 2009).

China's decision to adopt a semi-market economy resulted in "a break with China's long history of . . . subordinating law to the exercise of State power" (Wouters *et al.*, 2004, p. 247). This resulted in the adoption of the rule of law and the formulation of resource management frameworks where none had previously existed (Wouters *et al.*, 2004). This included the 1988 Water Law and its revision, the 2002 Water Law, which determines the administrative and regulatory arrangements for managing China's water resources.

China's Constitution provides that water resources are owned by the state, on behalf of the people. The 2002 Water Law provides a framework for the integrated management of water resources, including the establishment of water rights. Relevantly, it provides for both water planning and for the regulation of abstraction of water resources through a licensing system. At a policy level, the 11th Five-Year Plan (2006–10) sets the Chinese government's highest strategic goals. It includes a goal of establishing an initial water right allocation system and water right transfer system (State Council, 2006). China's administrative arrangements for water reflect the hierarchical nature of the country's political system. The 2002 Water Law gives the Ministry of Water Resources (MWR) primary responsibility for managing water. Under the MWR sit seven river basin authorities, one for each of China's major basins, as well as water resources departments and bureaux at provincial, prefecture and county levels. Lower levels of administration are subordinate to those above them, but superior laws and regulations will often leave discretion as to how they are to be implemented locally (Liu & Speed, 2009). The 2002 Water Law's water planning framework consists of a number of inter-related plans. This includes separate plans at both the basin and regional level, including both comprehensive plans, as well as a number of 'special' plans (e.g., hydropower development, flood management, irrigation and drainage which are subordinate to the comprehensive plans). Water in a trans-provincial basin is allocated by a basin allocation plan amongst different provinces. Provincial plans then allocate their share amongst prefectures, and prefectural plans allocate water amongst counties. The extent to which these arrangements have been implemented varies across the country (Gao, 2006; Shen & Speed, 2009). Water abstraction permits are required to take water from a watercourse or aquifer. These often operate as a bulk water entitlement, with the entitlement used to supply multiple water users within either an irrigation district or as part of an urban water supply system (Calow *et al.*, 2009; Cosier & Shen, 2009).

While there have been many phases in China's history of water resources management, the most recent chapters have been driven by water scarcity and water pollution issues, and associated conflicts over the limited available supplies. In particular, overextraction from the Yellow River—which resulted in the river ceasing to flow for prolonged periods during the 1980s and 1990s—led to the development of China's first major water allocation plan and annual water regulation plan. Experience in the Yellow River has been a major influencing factor in the development of China's new water management framework (Shen & Speed, 2009). The most recent management shift has been towards a rights-based approach to water management. This has involved a range of initiatives, including:

- implementation by basin authorities and provincial governments of the requirements of the 2002 Water Law, including through the adoption of improved water licensing systems and the establishment of water resources allocation plans;
- defining water entitlements at the farmer level within irrigation districts; and
- pilot water trading projects (WET, 2006; Gao, 2006).

Key Lessons from Water Rights Implementation

Apart from the institutional differences, the approaches to water rights in China and Australia are—not surprisingly—similar in terms of their basic structure. Likewise, many of the issues identified in Australia during the reform process are evident in China—incremental licensing, poorly specified entitlements, limited recognition of environmental flows, a lack of security of entitlement and restrictions on transferring water rights (Gao, 2006).

Australia has made significant progress in addressing many of these issues (Parker, 2006; Hamstead *et al.*, 2008). At the same time, the extended drought over recent years has highlighted the shortcomings of some of the approaches taken in implementing water rights. China still has some way to go in resolving these issues. While the 2002 Water Law provides a sound basis for a water rights system, there is a clear need to strengthen the interactions between different elements of the water rights framework. In particular, connections between allocation plans at different administrative levels and the abstraction licensing system are weak. Little consideration is given to environmental requirements for water and there are few protections against changes to rights. Together, these undermine the certainty, security and sustainability of the water rights system (WET, 2006; Xie, 2008).

A number of specific technical issues will need to be addressed to resolve these issues. This paper, however, focuses on the broader policy issues that have emerged and the approaches that have been adopted to address them. In particular, it looks at lessons from Australia, where a longer history of water rights implementation has given greater insights into the benefits and pitfalls of different policy approaches. At the same time, China's management of trans-provincial rivers offers important lessons.

Protection of Rights and Assignment of Risk

While a water rights system is fundamentally about granting users secure entitlements to water, there are significant issues concerning the appropriate level of security. These arise because of (1) the public nature of water resources and (2) the uncertainties of the science associated with allocating water in the first place. Despite best endeavours when initializing water rights, it may be necessary to later reduce at least some rights, either to ensure water is available for a critical public purpose (e.g. for domestic supplies), to provide more water for the environment (where science suggests more is necessary than originally thought) or because of reductions to the total water resource (e.g. due to climate change). Water rights systems need to balance these issues—i.e. the need for adaptive management—with users' requirements for security and certainty.

In China, water rights are not well protected. Water abstraction permits are governed by the 2004 Law of Administrative Licences, which provides that compensation is payable for the cancellation or change of a licence if the licensee experiences 'property losses'. It is not clear that this provision would provide any protection for a water abstraction permit holder in the case of modification (WET, 2006) and discussions with local water officials suggest that a much higher priority is placed on state control than on individual guarantees. Indeed, China's first property rights law was made only in 2007—with security only recently granted in respect of land title, it seems likely that it will be some time before similar guarantees are given for (less tangible) water rights.

In Australia, the NWI sets the principles for sharing the risk between government and water users for changes to water rights. It provides that government bears the risk for any adjustments during the life of a statutory planning cycle (e.g. 10 years). Water entitlement holders bear the risk for reductions of up to three per cent that are made at the end of the statutory planning period as a result of improved knowledge on the sustainable limits of a system. Any greater reductions will be at the risk of government (with an agreed formula for sharing the risk between the state and federal governments), although this commitment only applies in some states to plans made or varied after 2014. Water users bear the risk for changes to rights associated with climate change or periodic drought, while government bears the risk of changes associated with changes to government policy (NWI, 2004, clauses 48–50). The extent to which these principles have been given effect by state legislation varies. In New South Wales, the law generally provides for government to pay compensation for changes to entitlements in accordance with the risk-sharing principles described in the NWI (Water Management Act 2000 (NSW), sections 87-87A). In Queensland, a water entitlement holder is entitled to compensation for any change to a water resource plan during its 10-year life that reduces the value of their entitlement (Water Act 2000 (Qld), section 986).

As China further expands its water rights system, and particularly as it implements water trading, it is likely there will be calls for stronger protection of water rights. Experience in Australia suggests the following:

- The assignment of risk needs to be cognisant of the level of uncertainty that exists over future water availability. For example, governments should consider the merits of giving lesser guarantees where the available science or hydrological information is not sufficient to support confident projections of future needs and availability.
- Transparency in the decision-making process is critical. The law should clearly state the way and the circumstances in which water rights can be adjusted (regardless of whether compensation will be paid), and comprehensive monitoring and regular reporting (e.g. on water availability and a river's environmental condition) should provide the information necessary for water users to assess the likelihood of future changes to their rights.

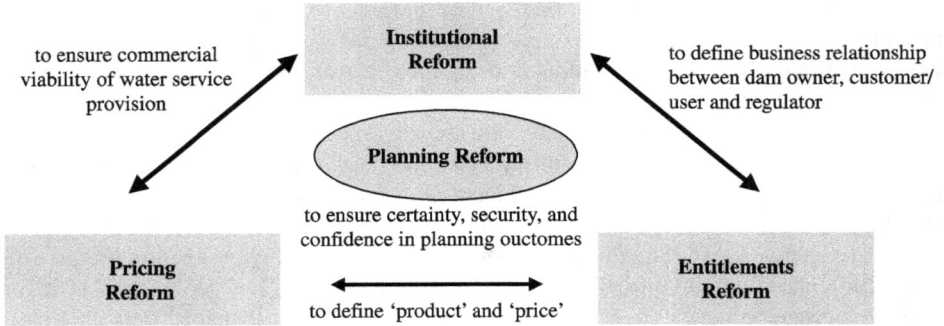

Figure 2. An integrated approach to water reform. *Source:* Parker (2006).

- Defining rights (particularly where they have not been defined before) can in itself provide a level of protection to users, even where there is no legal guarantee of compensation. Issuing a piece of paper to a water user can create an expectation and result in political pressure to protect the 'right', regardless of any obligation at law. This can have both positive consequences (e.g. providing confidence to users) as well as negative ones (see below).
- Regardless of what may be possible under law, reducing a user's water rights is generally a very difficult thing to do politically. For example, Australia is buying water entitlements from irrigators in the Murray-Darling on a voluntary basis, even though there is legal scope to reduce total water abstractions without paying compensation[2] and despite the risk-sharing arrangements set out in the NWI. The economic and social dependencies that have flowed from the grant of water rights (at the state level) has created major political challenges for government (at both the federal and state level) in terms of its options for reducing total water use. Ultimately, in the eyes of many in the community, it is not water users but government—as the one responsible for managing the resource—that is seen as responsible for bearing the cost of any reductions to water rights.

Integrating Water Rights with Other Management Initiatives

While the term 'integrated' water management is used in many contexts, the development of water rights in Australia and China provide two examples of the types of integration that can be necessary.

In Australia, creating a water rights system has required both planning and entitlements reform (i.e. reforms to the way water resources are shared and individual water rights are specified) as well as parallel institutional and pricing reforms. Planning reform has been necessary to better quantify the resource and share it between the environment and users to provide transparency and greater certainty. Institutional reforms have been necessary to separate the roles of regulator, resource manager and service provider. At the same time, pricing reforms have been necessary to ensure that newly created service providers are commercially viable (Parker, 2006; Claydon & Onta, 2008). Each of these elements is related to the others. The water rights of new service providers needed to be defined (historically the state had not granted itself a water right), as did the relationship between the service provider and the water right holder—in terms of what 'product' they were being supplied, and what price they would have to pay for it. These relationships are shown in Figure 2. The interdependencies highlight that reforms of this kind need to be undertaken as an integrated package, and cannot be done in isolation. It is also desirable for these reforms to have been undertaken before water markets start to fully operate.

Quite separate from these issues is the integration of water rights with other water (or water-related) management activities. China's 2002 Water Law defines the relationship between a range of plans, including water allocation plans, water quality protection plans, water supply plans, flood management plans, and plans for water and soil conservation. It also requires a number of 'comprehensive' plans—at the national, basin and regional level—which further define how the different aspects of water resources management will relate to one another (Shen & Speed, 2009).

There is recognition in Australia of the need to better coordinate water allocation planning with catchment management plans, regional economic development and/or growth management plans, and water supply plans (Hamstead *et al.*, 2008). In particular, extended drought has placed pressure on water managers to better coordinate water allocation plans and water supply planning (e.g. QWC, 2007). While overall there remains a need for improved coordination of the different aspects of the water management system in Australia, the Water Act 2007, in providing for a water plan for the Murray-Darling Basin, should go some way to addressing these issues in that basin.

Environmental Flows and Water Rights

Water rights systems, in capping the total water available for abstraction, will—whether by design or by default—also determine the water that is made available for the in-stream environment. Environmental flows are increasingly recognized as essential to the well-being of river systems (Bunn & Arthington, 2002) and this is reflected in high-level policies in both Australia and China. However, the success of either country in providing water for the environment has been mixed at best (WET, 2006; Connell, 2007; Wang *et al.*, 2009).

In China, water is often not explicitly set aside for the environment. Where it is, targets are often set for a singular purpose (e.g. sediment transport) or have been based on simplistic assessment methods. While more sophisticated techniques have been developed in recent years, these have not yet been put in to practice (Wang *et al.*, 2009). In Australia, advice about the necessary environmental flows required to meet a range of environmental objectives and outcomes has to date been typically provided by an expert panel as part of the water resource planning process undertaken at the state level. That advice is then considered by government in the broader planning context. Plans usually set environmental objectives, and then aim to achieve those by capping abstractions, setting operational rules for water infrastructure, and (in some cases) granting 'environmental allocations' and designating a manager to utilize the allocation for the benefit of the environment. Water resources plans, incorporating some form of environmental flow requirements, are now in place for most major catchments (Hamstead *et al.*, 2008).

China is in the process of amending the comprehensive plans for its major river basins to include environmental flow requirements for the first time. Australia's efforts in this field to date suggest the following issues should be considered:

- It can take many years to build the scientific base to support environmental flow assessments. Research needs to be structured to support future management requirements. However, a lack of information should not deter decision-making. Experience shows that the longer problems are left, the harder they are to reverse.
- Research has shown that in the case of environmental flows, it is possible to do 'more with less'. The frequency and timing of flows can be ecologically as important as the volume. As such, it is not always necessary to reduce existing water rights to improve environmental conditions: river health can sometimes be improved through better management of the water already allocated for environmental purposes (WET, 2007).
- Allocating water rights involves a trade-off between consumptive requirements and environmental needs. Where a government decides to reduce (or not to increase) the flows available to the environment, the decision to do so should be transparent. It is likely that at times other interests will take priority over the environment. However, this should occur in an informed and public manner, to allow all involved to understand the consequences of the decision (Hamstead *et al.*, 2008).
- The dire situation in the Murray-Darling Basin suggests that the major efforts by both the federal and state governments over the past decade to provide water for the environment have not yet been effective in restoring flows to a sustainable level. Generally, water resource plans in Australia have sought to 'maintain current environmental values', rather than improve the ecological condition (Hamstead *et al.*, 2008). Some commentators have argued that abstraction levels have been set based on what can be spared from current use rather than what is required to achieve sustainability (Connell, 2007). As such, the political decision has often been to set abstraction caps based on existing levels of use. In over-allocated systems such as the Murray-Darling, this has proved insufficient. Australia is now facing the consequences of that approach, including the AU\$3.1 billion cost of buying back entitlements to return water to the environment. Amongst other things, this suggests (1) that employing sophisticated assessment methodologies will not solve ecological problems without adequate political will and (2) providing (or not providing) adequate water for the environment comes at a cost, either now or in the future.

Trans-Regional Approach to Water Rights

The woes of the Yellow River and the Murray-Darling Basin have been central to changes in water resources management in their respective countries. Both are notable for spanning multiple jurisdictions (the Yellow River crosses 10 provinces and autonomous regions, while the Murray-Darling Basin spans four states and one territory) and both have been the subject of significant conflict over water sharing between jurisdictions.

However, to date, China's hierarchical administrative system has given it the capacity to better address these conflicts. China's State Council first approved a water allocation plan for the Yellow River in 1987, defining the shares of each of the 10 provinces to the 58 billion m^3 available (on average) each year. The Yellow River Conservancy Commission has overall responsibility for water within the basin. It manages available supplies through an annual regulation plan (defining the shares of each province for the year), adjusting the plan on a monthly and ten-day basis, to achieve trans-provincial flow requirements. The Commission owns and operates metering and monitoring systems to ensure compliance, and is responsible for operation of some of the major infrastructure on the river. Below the basin plan sit regional plans to allocate the water amongst sub-regions (Shen & Speed, 2009). In the Murray-Darling, in contrast, the division of responsibilities between the states has meant that the various state water-sharing arrangements in place over the past century have been reached by consensus, rather than dictated by the Federal Government. This has not enabled a whole of basin water plan to be developed and agreed. The limitations (and failings) of this approach are evidenced by the decision in 2008 of the states to refer certain powers over water planning and management in the Murray-Darling to the Federal Government.

The Chinese approach to trans-provincial water management should provide a valuable guide for Australian policy makers and the new Murray-Darling Basin Authority as they develop the Murray-Darling's first whole-of-basin water plan. This will include how best to establish the connections between the basin plan and (state-prepared) catchment plans, how to determine annual allocations based on actual availability, approaches to specifying environmental flow requirements and responsibilities for monitoring and enforcement.

Trading and Reallocating Water Rights

Both Australia and China have experienced major water shortages and the challenge of providing water for development in fully allocated systems. In overcoming this challenge, the two have adopted different approaches to transferring water rights, which reflect their different political and cultural contexts.

Water rights are readily tradable in a number of parts of Australia. Water users can sell, in some cases via online systems, their water allocation (the water available to them that year) or their water access entitlement (their long-term right to water). Trading must comply with rules established to protect against adverse environmental or third party impacts (Parker, 2006). The systems in place have allowed farmers to respond to seasonal variations (through temporary trading of annual allocations) as well as to make longer-term structural adjustments. Evidence suggests that there has been significant economic benefit from water trading (NWC, 2008).

Water trading in China is at an elementary stage and is generally occurring outside of a structured trading framework (WET, 2006). Examples of water trades to date include sales from one county government to another, and the transfer of water rights from irrigation districts to industries following water efficiency initiatives (Gao, 2006; Speed, 2009). Notably, these have been driven by government and not the free market.

Emerging lessons from water trading in Australia and China include:

- As with any rights-based system, water rights need to be clearly specified before trading occurs, to define what is being transferred and to protect all parties against adverse impacts of the transfer (Productivity Commission, 2003).
- Temporary trading—the sale of some or all of the water available in a particular year—dominates the Australian water market, accounting for around two-thirds of water trades (NWC, 2008). This suggests that most water is traded to accommodate changes in requirements during a given year, rather than because of structural adjustments. China's current reallocation system does yet not provide any scope for temporary trades.

- There have been concerns in both countries of the possible social consequences from water reallocations (Speed, 2009; Parker, 2006). However, where when and if necessary, social or sectoral interests can be protected by trading rules that limit the volume of water that can shift away from a region or a sector (Speed, 2009; Parker, 2006) provided there are legitimate and transparent reasons to regulate the market in this way. These can, of course, be adjusted over time if it becomes apparent that concerns over the impact of water transfers are not materializing (or, alternatively, if new concerns arise).[3]
- The high level of government involvement in water transfer projects in China— while in conflict with the concept of pure market-based approach to allocating water—has advantages. First, it has allowed transfers to be made based on whole-of-government priorities (which will not necessarily align with short-term economic benefits). Secondly, it has provided for economies of scale in undertaking water efficiency measures, such as the lining of irrigation channels (WET, 2007). Increased government involvement can also be justified in cases of ensuring critical human need and urban water security are met (QWC, 2007).
- Low cost trading requires water rights to be separated from other approvals—such as works approvals, water use conditions and contractual obligations. Historically these have been bundled in a single authorization. As it is only the water right that is traded, the right needs to be free from other encumbrances. Otherwise, there is likely to be a high transaction cost associated with transferring the right (WET, 2007).

Conclusions

The trends in water management in both countries show interesting similarities and contrasts. In Australia, there is a significant shift towards federal control of water resources, most obviously in the stressed Murray-Darling Basin, coupled with a market-based approach to the re-allocation of water. At the same time, the Federal Government's \$3.1 billion water buy-back scheme in the Murray-Darling means that it will almost certainly become the largest buyer in Australia's water markets. Meanwhile, in China, the government has somewhat loosened its tight grip and there are efforts to decentralize some of the roles of water management agencies. This is particularly evident at the irrigation district level, where management responsibilities are shifting over to 'water user associations' (Calow *et al.*, 2009). In both countries, clarifying the roles of basin authorities—and generally strengthening river basin management—has become a major priority.

The development, in both Australia and China, of rights-based systems for managing water resources has provided experiences from each country that would be of benefit for the other. The issue of compensation for changes to entitlements—however decided—will need to be addressed in China if water users are to be given some level of protection. It may be that security initially comes in the form of procedural guarantees, rather than financial ones, given political sentiment.

In terms of trading of rights, Australia's experience with temporary trading suggests that China would benefit from allowing seasonal reallocations, not just permanent transfers of water rights. At the same time, the Chinese government's hands-on role in water reallocations is in stark contrast to the Australian approach. Australian water managers would do well to note the approach adopted in China, and particularly in the Yellow River, to sharing water in a trans-regional river basin. China's hierarchical political system has left it better placed to establish the regulations and institutions necessary to share water between the different provinces. It is possible that the new management arrangements being developed for the Murray-Darling Basin will ultimately need to follow a similar approach to those already in place in China, noting the long-standing tensions that have existed between the states when it comes to sharing Murray-Darling waters.

The types of water rights systems that have evolved in the two countries are as much a reflection of political and cultural differences as they are related to different water issues. China has opted for a higher level of government control, whereas in Australia market forces have a greater role. Ultimately, for a given country, deciding the right point to settle on the water rights spectrum will depend on the country's particular political circumstances, the level of sophistication and capacity within the relevant management agencies, and its overall policy objectives.

Acknowledgements

This paper is the result of a project undertaken under the auspices of the Australian Department of the Environment, Water, Heritage and the Arts and the Chinese Ministry of Water Resources, with funding provided by AusAID, the Australian Agency for International Development.

Notes

1. In fact, in Australia the term 'water right' is seldom used, and has been deliberately avoided in legislation to remove some of the perceptions associated with the term. Instead, terms such as 'water access entitlements' (the long-term entitlement) or 'water allocation' (an annual volume of water available to a user) are typically used to refer to interests in water.
2. State governments are not bound by any constitutional requirement to pay compensation in the event of the acquisition of property (see *New South Wales v Commonwealth* (1915) 20 CLR 54). It is legally debatable whether an action by the Commonwealth—for example through making a basin plan—that resulted in change to a water entitlement (granted by a state government) would amount to the acquisition of property for the purposes of Section 51(xxxi) of the Commonwealth Constitution, and thus require compensation.
3. By way of example, in Queensland, legislation allows for water trading rules to place restrictions on the water trading from one sector to another. This was included to address concerns that water would shift away from particular sectors (e.g. agriculture) and have adverse social consequences. In practice, when it has come to prepare water trading rules as part of the water planning process, there has not been support at the community level for imposing these kinds of restrictions.

References

ABS (2007) *Water and the Murray-Darling Basin: A Statistical Profile, 2000–01 to 2005–06* (Canberra: Australian Bureau of Statistics). Available at: http://www.abs.gov.au/ausstats/abs@.nsf/mf/4610.0.55.007 (accessed 1 December 2008).

Banyard, R. & Kwaymullina, A. (2000) Tradable water rights implementation in Western Australia, *Environmental and Planning Law Journal*, 17(4), pp. 315–333.

Bunn, S. E. & Arthington, A. H. (2002) Basic principles and ecological consequences of altered flow regimes for aquatic biodiversity, *Environmental Management*, 30, pp. 492–507.

Calow, R. C., Howarth, S. E. & Wang, J. (2009) Irrigation development and water rights reform in China, *International Journal of Water Resources Development*, 25(2), pp. 227–248.

Claydon, G. K. & Onta, P. S. (2009) Smart water planning in Queensland, Australia, paper presented at *OzWater'09 Conference*, Melbourne, Australia, 16–18 March 2009.

COAG (Council of Australian Governments) (1993) *Report from the Working Group on Water Resources Policy* (Canberra: Council of Australian Governments).

COAG (Council of Australian Governments) (2004) *Intergovernmental agreement on a National Water Initiative*, June 25 (Canberra: Council of Australian Governments).

Connell, D. (2007) The sustainability of sustainable limits to extractions informing the NWI, Canberra: Land and Water Australia, unpublished report, cited by Hamstead *et al.* (2008).

Constitution of the People's Republic of China (adopted 4 December 1982). Available at: http://english.peopledaily.com.cn/constitution/constitution.html (unofficial translation) (accessed 30 November 2008).

Cosier, M. & Shen, D. (2009) Urban water management in China, *International Journal of Water Resources Development*, 25(2), pp. 249–268.

CSIRO (2008) *Water Availability in the Murray-Darling Basin*, A Report to the Australian Government from the CSIRO Murray-Darling Basin Sustainable Yields Project, CSIRO, Australia. Available at: http://www.csiro.au/files/files/pn9o.pdf (accessed 5 December 2008).

DEWHA (Australian Department of Environment, Water, Heritage and the Arts) (2008) Rudd government to invest $12.9 billion in water, Media release, Canberra. Available at: http://www.environment.gov.au/minister/wong/2008/pubs/mr20080429.pdf (accessed 30 November 2008).

Gao, E. (2006) *Water Rights System Development in China* [in Chinese] (Beijing: China Water and Hydropower Publishing).

Hamstead, M., Baldwin, C. & O'Keefe, V. (2008) *Water Allocation Planning in Australia: Current Practices and Lessons Learned*, Waterlines Occasional Paper No. 6, April, Canberra: National Water Commission.

Liu, B. & Speed, R. (2009) Water resources management in the People's Republic of China, *International Journal of Water Resources Development*, 25(2), pp. 193–208.

MWR (Ministry of Water Resources) (2007) *Annual Water Statistics (Beijing)*. Available at: http://www.mwr.gov.cn/english/ (accessed 20 November 2008).

NCC (1994) *Council of Australian Governments Agreement on Water Resources Policy* (Canberra: National Competition Council).

NWC (2005) *Australian Water Resources 2005* (Canberra: National Water Commission). Available at: http://www.water.gov.au/Default.aspx (accessed 1 December 2008).

NWC (2008) *Australian Water Markets Report 2007–2008* (Canberra: NWC). Available at: http://www.nwc.gov.au/resources/documents/AWMR2007-08COMPLETE.pdf (accessed 19 May 2009).

NWI (Intergovernmental Agreement on a National Water Initiative) (2004) *Agreement of the Council of Australian Governments*. Available at: http://www.nwc.gov.au/www/html/117-national-water-initiative.asp (accessed 28 April 2009).

OECD (2007) *OECD Environmental Performance Reviews: China* (Paris: OECD Publishing).

Parker, S. (2006) Market mechanisms in water allocation in Australia, in: *Proceedings of OECD Workshop on Environment, Resources and Agricultural Policies in China*, June 19–21, Beijing, China.

Productivity Commission (2003) Water rights arrangements in Australia and overseas. Commission Research Paper, Melbourne: Productivity Commission.

Property Rights Law of People's Republic of China (2007) National People's Congress. Available at: http://www.gov.cn/flfg/2007-03/19/content_554452.htm (accessed 30 November 2008).

QWC (Queensland Water Commission) (2007) *Urban Water Supply Arrangements in SEQ*, Queensland Water Commission's Final Report to the Queensland Government, May. Available at: http://www.qwc.qld.gov.au/Urban+Water+Supply+Arrangements+Report (accessed 30 November 2008).

SBS (Special Broadcasting Service) (2006) *SBS World Guide*, 14th ed (Prahran: Hardie Grant Books).

Shen, D. (2004) The 2002 Water Law: Its impacts on river basin management in China, *Water Policy*, 6, pp. 345–364.

Shen, D. & Speed, R. (2009) Water resources allocation in the People's Republic of China, *International Journal of Water Resources Development*, 25(2), pp. 209–225.

Speed, R. (2009) Transferring and trading water rights in the People's Republic of China, *International Journal of Water Resources Development*, 25(2), pp. 269–281.

State Council (2006) *National Economic and Social Development Plan for Eleventh Five-Year Period, 2006–2010*. Available at: http://english.gov.cn/special/115y_fd.htm (accessed 15 November 2008).

Sun, X. (2002) *Comprehensive Report of Strategy on Water Resources for China's Sustainable Development* (Beijing: The Editorial Group).

Wang, X., Zhang, Y. & James, C. (2009) Approaches to providing and managing environmental flows in China, *International Journal of Water Resources Development*, 25(2), pp.

Water Act 1989 (Victoria). Available at: http://www.legislation.vic.gov.au/ (accessed 19 May 2009).

Water Act 2000 (Queensland). Available at: http://www.legislation.qld.gov.au/Acts_SLs/Acts_SL_W.htm (accessed 19 May 2009).

Water Law (2002) National People's Congress of the People's Republic of China.

Water Law (2007) Commonwealth of Australia. Available at: http://www.comlaw.gov.au (accessed 30 October 2008).

Water Management Act 2000 (New South Wales). Available at: http://www.legislation.nsw.gov.au/maintop/scanact/inforce/NONE/0 (accessed 19 May 2009).

WET (2006) *Water Entitlements and Trading Project (WET Phase 1) Final Report* November 2006 [in English and Chinese] (Beijing: Ministry of Water Resources, People's Republic of China and Canberra: Department of Agriculture, Fisheries and Forestry, Australian Government). Available at: http://www.environment.gov.au/water/action/international/wet1.html.

WET (2007) *Water Entitlements and Trading Project (WET Phase 2) Final Report* December 2007 [in English and Chinese] (Beijing: Ministry of Water Resources, People's Republic of China and Canberra: Department of the Environment, Water, Heritage and the Arts, Australian Government). Available at: http://www.environment.gov.au/water/action/international/wet2.html

Wouters, P., Hu, D., Zhang, J., Tarlock, D. & Andrews-Speed, P. (2004) The new development of water law in China, *University of Denver Water Law Review*, 7(2), pp. 243–308.

Xie, J. (2008) *Addressing China's Water Scarcity: A Synthesis of Recommendations for Selected Water Resource Management Issues* (Washington, DC: World Bank Publications).

Yellow River Commission (n.d.) *About the Yellow River.* Available at: http://www.yrcc.gov.cn/eng/about_yr/jj_13104025172.html (accessed 1 December 2008).

Index

Page numbers in **bold** refer to figures. Page numbers in *italics* refer to tables or boxes.

abstractor rights 83, 84
Adelaide 206
agricultural water management 3, 17, 39–57, 89
allocation of water rights 3
annual allocation of water 26
Annual Regulation Plans 46ff.
assignment of risk 208–10
Australia 2, 22, 62, 82, 83, 91, 126, 150: comparison with China 3, 201–15, *203*, *204*, **210**; water reform 205–7
Average Flow in the Lowest Flow Season method 98, *109*

Baiyangdian Lake 101
Baizhao gauge **116**, 131, **131**, 133
balancing environmental flows with water supply reliability 143–65
bankfull 126, **126**, *127*, 130, 131, 134, *135–6*, *152*, 153ff., **153**, **158**
Baotou City 185, **186**
baseflow 130, **130**, 131, 137, 155
Bashang Region *42*, 50, 51, 54
basic food security 170
Beijing 44, 62, **63**, 66–9, **68**, 76ff.
Beijing Drainage Group Company 67ff., **68**
Beijing Municipal Government 66ff., **68**
Beijing Water Saving Management Centre **68**, 69
Beijing Waterworks Group Company 67ff., **68**, 76
benchmarking methodology 118

Black River *see* Hei River
Boisten Lake 100
bottom-up methodologies 118, 121
building-block methodology (BBM) 118, 121

cash cropping *42*
Changtan Reservoir **116**, 117, 145ff., **147**, 161ff., **161**
Channel Scheduling Plan 32
Chaobai River 66
Chile 22, 82
Cixi County 74, 85
Commission on Dams 95
comparison of water rights systems in China and Australia 3, 201–15: China and Australia compared and contrasted 203–5, *203*, *204*; conclusions 214–15; key lessons from water rights implementation 208–14, **210**; nature and theory of water rights 202–3; trans-regional approach to water rights 212–13; water reform 205–8
Constitution of the People's Republic of China **11**, 207
continuum of water rights *203*
Contract Law 1999
Contribution of Channel Divisions method 99, *110*
Council of Australian Governments 206
Cultural Revolution 41

Dadu River *109*

Daling River 27
dams 103
Darling River *204*
Datong City 44
defining rights 210
Deng Xiaoping 41
development of a water rights system 1–4
Dongjuyan Lake 100–101
Dongyang County 13, 85, 88
downstream response to imposed flow
    transformation (DRIFT) process 119
drought contingency plans 23

Ecological Hydraulic method 98, *111*
Ecological Limits of Hydraulic Alteration
    (ELOHA) 120–21
ecological water allocation 172
efficiency in water use 169
effency principle 170
end users: rights 83; water supply to 23,
    25–6
environmental flow component (EFC)
    119–20
environmental flows 3, 34: and water
    rights 211–12; application in water
    resources management 99–100, **100**;
    asset-based, holistic approach 113–39;
    balancing with water supply reliability
    143–65; brief history 96–8;
    deficiencies in methodologies 101–3;
    Jiao River Basin case study 113–39,
    116, *123*, *124*, **126**, *126–9*, **130**, **131**,
    132, *135–6*; obstacles to
    implementation 102–3; provision and
    management 95–104; review of
    methodologies 98–101, **98**, *109–12*;
    Taizhou case study 143–65, **146**, **147**,
    149, **150**, *152*, **153**, **156**, **157**, **158**,
    160, **161**
environmental safety 16–17
estuarine flows 137–8

fairness in water allocation 168–9
fairness principle 170
Fang Creek **116**, *124*, **147**, *152*, **157**, 158
Federal Energy Regulatory Commission
    (USA) 103

Federal Government (Australia) 212
fish diversity and abundance 126, *128–9*
fisheries maintenance 126, *128–9*
Five-Year Plan, 10th 14
Five-Year Plan, 11th 1–2, 14, 43, 54,
    113, 207
flood control 7, 9, 14
Flood Control Law 1997 9–11, **11**
FLOWS method 121–38, *123*, *124*, **126**,
    127–9, **130**, *135–6*,
France 62, 71
Fuhe River 27
Fujian Province 27, 84

Gansu Province 26, 49, 51, 52, 87,
    173–85, **173**, **174**
genetic algorithm (GA) 169
geomorphic forms 126, *127*
government-facilitated reallocations 202
groundwater **7**, *8*, 9
groundwater allocation 172–3
groundwater development 41–2, *42*
groundwater rights 50–54
group water entitlements (GWE) 190ff.
Guidelines on Technical Management of
    Irrigation and Drainage Projects
    1999 26
Gulang County 173, **174**, *175–6*, **177**,
    178–9, **180–182**

habitat simulation methods of
    environmental flow 98–9, **98**, *111–12*
Hai River 9, 99, 102, *109*, *112*
Hai River Basin 6, **7**, *8*, *10*, 49, 66, 99,
    100, 101
Han River *111*
Hangjin County 186
Hangjin Irrigation District 3, 17, 31–2, 48,
    49, 86, 89, 90: water rights
    constitution 185–90, **186**, **189**
Hangjin Irrigation Management Bureau
    (HIMB) 187, 190, *193*, 194
Hebei Province 27, 41, *42*, 50ff., 66, *188*
Hei River 26ff., 101
Hei River Basin 49, 87
Heilong River 99, *111*
Henan Province 41, *188*

Hengjin Reservoir 85, 88, 89
Hetao Irrigation District *44*, 48
Hexi Corridor 12
high flow *127–9*, *135–6*, *152*, 153ff., **153**, **158**
Hohhot City 185, **186**
holistic methods of environmental flow 98–9, **98**, *112*, 120–21
Household Responsibility System 41
Huai River 100, 102, *109–12*
Huai River Basin 6, **7**, *8*, *10*, 40, 99ff., **100**
Huairou Reservoir 66, 67
Huang River Basin **100**
Huangpu River 70
Hui Autonomous Region of Ningxia 85–7, *188*
Hun River *109*, *111*
Hunan Province 43
hydraulic methods of environmental flow 98–9, **98**, *110–11*, 120, *123*
hydraulic modelling and hydrological analysis 130–33, **130**, **131**, **132**
hydrological methods of environmental flow 98–9, **98**, *109–10*, 119–20, *123*

Indicators of Hydrological Alteration (IHA) 119–20, 138–9
Inner Mongolia Autonomous Region 26, 27, 44, 47ff., 85, 86, 185–90, **186**, *188*, *189*, **189**
Inner Mongolia Yellow River Irrigation Management Bureau (IM-YRIMB) 48, 187, *189*
Integrated Quality and Quantity Model (IQQM) 125, 133, 137, 148, **150**, 151
integrated water resources management (IWRM) 40, 54, 169
integrating water reform **210**
integrating water rights with other management initiatives 210–11
intelligent cards (ICs) 44, 51, 53, 54
irrigation development 3, 39–57: conclusions and recommendations 54–7; groundwater rights 50–54; history 40–42; management challenges 40–47, *42*, *44*; reform of water rights 45–7; responses to challenges 43–5;

rights management in practice 47–54
irrigation districts 25–6, 40ff., 87, 186: agencies 44
Irrigation Water Scheduling Plan 32

Jiangsu Province 70
Jiangxi Province 27
Jiao River *112*, **116**, **147**, 161
Jiao River Basin 144ff.
Jiao River Basin case study 3, 113–39: application of the FLOWS method 124–38; developing conceptual models 125, **126**; developing flow rules 134–8, *135–6*; discussion and conclusion 138–9; hydraulic modelling and hydrological analysis 130–33, **130**, **131**, **132**; identifying ecological assets 125; selecting an appropriate methodology 115–21; selecting representative reaches **116**, 124–5, *124*; setting management objectives 126–30, *127–9*; steps involved in FLOWS method 121–38, *123*, *124*, **126**, *127–9*, *135–6*; study reaches **116**; *see also* Taizhou case study
Jiaojiang District **116**, 145, **147**
Jin River 27, 30, *30*, 84
Jinchuan District 173, **174**, *175–6*, **177**, *178–9*, **180–182**
Jinhua Prefecture 85
Jinji Gate 30

Kangbao County *42*
key elements of water rights systems 202–3

Law of Administrative Licences 2004 209
Law on the Prevention and Control of Water Pollution 9, **11**, 66
Liangzhou District 173, **174**, *175–6*, **177**, *178–9*, **180–182**
Liao River 99, 102, *109ff.*
Liao River Basin **7**, *8*, *10*, 99
Liaoning Province 27
Ling River **116**, 118, 132, **147**, 161
Linhai City **116**, 132, **132**, **147**
Lishimen Reservoir **116**, 117, 145, **147**

Liyuan Irrigation District 49, 87
local rules and regulations **11**
low flow **126**, *127–9*, **130**, *135–6*, *152*,
    153ff., **153**, **156**, **158**

major river basins **7**, *8*
Millennium Development Goals 16, 61
Min River 41, 42
Minimum Flow in Dry Season method
    98–9, *109*
Ministry of Environmental Protection
    (MEP) 11, 66, 76, 96
Ministry of Health 65
Ministry of Housing and Urban-Rural
    Development 66
Ministry of Water Resources (MWR) 2, 9,
    11–12, 22, 43, 45, 49, 63, 66, 76, 87,
    97, 207
Minqin County 51, 54, 173, **174**, *175–6*,
    177, *178–9*, 180ff., **180–82**
Miyun Reservoir 66, 67
Montana method *see* Tennant method
Monthly Guarantee Rate method 98, *109*
Murray–Darling Basin *204*, 205ff.
Murray–Darling Basin Authority 206
Murray River *204*, 20-6
Murrumbidgee River *204*

National Development and Reform
    Commission 65
National Plan for Building a Water-Saving
    Society 14
National Water Initiative (NWI) 206, 209
national water policy issues 2–3
National Water Savings Office 44
Nature Conservancy 119ff., 138
nature of water rights 203
New South Wales 206, 209
Ningbo Prefecture 74
Ningxia, Hui Autonomous Region of
    85–7, *188*
Ningxia Yellow River Water Rights
    Conversion Master Plan 86–7
Nioutoushan Reservoir **116**, 117, 145, **147**
North China Plain 9, 42, 45, 49, 50
North–South Water Transfer Project 66, 67
Northern Silk Road 12

Northwest River Basin **7**, *8*, *10*

objective function 169
Ordos City 185, 186, **186**
overbank flows 126, *127–9*, 134, *135–6*,
    152, 153ff., **153**, **158**

Pearl River Basin 6, **7**, 8, *10*
People's Republic of China (PRC) 41, *44*
planning mechanism for transferring water
    rights 82
pollution 7, 9, 13, 15, 16, 102–3, 205
practical application of water resources
    allocation 26–32, *29*, *30*
pricing mechanisms 202
protection of rights 208–10
provincial water resources departments 12
pulses (river flow) 126, *127–9*, 134,
    135–6, *152*, 153ff., **153**, **158**

Qiangtang River 27, 73
Qing Dynasty *44*
Qinghai Province 49, *188*
Qinghai-Tibet Plateau 6
quantity and quality control mechanisms **15**
Quanzhou Prefecture 27, 30
Queensland 206, 209
quotas 15, **16**

range of variability approach (RVA) 119
reducing water abstractions 202
regional water rights 83–4, 85
Regulations on the Administration of
    Water Abstraction Licensing and
    Collection of Water Resources
    Charges 2006 63–5
Regulations on Urban Water Saving
    Management 1988 65, 69
regulatory system for transferring water
    rights 82
reliability of water supplies 159–60,
    160, **161**
    balancing with environmental flows
    143–65
Republic of China 41
reservoirs 103
rights management in practice 47–54

riparian vegetation 126, *127*
river basin and regional water resources
    allocation 23–4
river basins, major **7**, *8*, *10*
river pollution 102–3
river systems 95

Sanmenxia Reservoir 103
saving and transferring water 85–7
scarcity of water 7, 9, 14, 205
sediment transport methods of
    environmental flow 98–9, **98**, *112*
Shandong Province 41, *188*
Shanghai 31, 62, **63**, 70–73, **71**, 76–7
Shanghai Municipal Government 70
Shanghai Pudong Veolia Water
    Company 71
Shangyu County 73, 74
Shanxi Province 41, *188*
Shanyi County *42*
Shaoxing County 73, 74, 85
Shaoxing Prefecture 62, **63**, 73–5
Shaoxing Water Affairs Corporation 75
Shawo River *111*
Shields Critical Shear Stress method
    130–31
Shifeng Creek **116**, 124, 132, 133, **147**,
    152, 156, **157**, 158
Shisandu Dam **116**
Shisandu Reservoir **116**, 117, 145, **147**,
    149, *149*, 156, **157**, **158**, 162
Shiyang River 26, 28
Shiyang River Basin 3, 53, 173ff., **173**
Shiyang River Basin case study 173–85:
    basing water allocation on current use
    174, *175*, **177**; basing water allocation
    on future demand 174–7, *176*, **177**,
    178-9, **180**; comparison and
    discussion 177–81, **180–82**; *see also*
    water rights allocation
Shiyang River Basin Restoration Plan
    173ff., **180–81**
Sichuan Province 41, *188*
Singapore 62
Song Dynasty 41
Songhua River 99, *109–11*
Songhua River Basin **7**, *8*, *10*

South Africa 22, 61, 119
South Australia 206
Southeast River Basin **7**, *8*, *10*
Southwest River Basin **7**, *8*, *10*
State Council 1, 2, 11, **11**, 22, 23, 28, 45,
    46, 49, 65, 85, 212
sub-optimal environmental flows regime
    151–4, *152*, **153**, 162–3
surface water *8*
surface water allocation 172–3
sustainable development 168

Tai Lake 101
Tai Lake Basin 70, 71
Taiwan 119
Taiyuan City 44
Taizhou case study 143–65: background
    145–8, **146**, **147**; compliance with
    environmental flow requirements
    154–9, **156**, **157**, **158**; conclusion
    163–5; cumulative impacts of
    development 160–61; exploring less
    intense options 161–2; methodology
    148–55; reliability of water supply
    148, 159–60, **160**, **161**; results
    156–63, **157**, **158**, **160**, **161**; study
    reaches **147**; sub-optimal
    environmental flows regime 151–4,
    152, **153**, 162–3; water resources
    allocation plan scenarios 148–50, *149*,
    150; *see also* Jiao River case study
Taizhou Comprehensive Plan for Water
    Resources 149, *149*, 161
Taizhou Prefecture 114ff., **114**
Taizi River *109*, *111*
Tang Dynasty 41
Tangpu Reservoir 73–4
Tarim River 27ff., 96, 97, 100, *111*, *112*
Tarim River Basin 96
temporary trading of water rights 83, 91
Tennant method 98, 103, *109*, 119, 138
Tennessee Valley Authority 103
theory of water rights 202–3
Tiancunshan Water Treatment Plant 67
Tianjin City 44, *188*
tide heights at Linhai gauge 132, **132**
top-down methodologies 118

transferring and trading water rights 3, 81–91, 203, 213–14: expanding markets 90; farmer-level transfers 90; fundamentals of a system 82–3; importance of well-defined rights 88; opportunities and challenges 87–91; pilot projects 85–7; policies and laws 83–4; temporary trading 83, 91; third party impacts 89; transfer framework 83–4; water sharing rules 88–9
transparency in decision-making 209
trans-regional approach to water rights 212–13

United Kingdom 62: Department for International Development (DFID) 52
United States 22, 62, 82, 103
Urban Planning Law 1989 65
urban water management 3, 25–6, 61–79: general framework 62–6, **64**; implementation of national policies and regulations 77–8; in Beijing 66–9, **68**; in Shanghai 70–73, **71**; in Shaoxing Prefecture 73–5; institutional arrangements 66–8, **68**, 70–71, **71**, 74; structure of management framework 75–7; supply and management practices 68–9, 71–3, 74–5
Urban Water Supply Quality Regulations 1994 65, 66, 77
user-level rights 83

vegetation, maintenance of 126, *127*
Victoria 206

water abstractors, regional supply to 23, 24–5
Water Act 1989 (Victoria) 206
Water Act 2000 (Queensland) 206, 209
Water Act 2007 (Commonwealth) 207
water affairs bureaus (WABs) 12, 31, 45–6, 50, 52, 74: Beijing 66ff., 76; Shanghai 70ff., **71**, 77
water balance constraint 171
water certificates 191–3, *192*, **193**
Water Conservation Project 52

water demand guarantee 168
Water Entitlements and Trading (WET) Project 2, 22, 67, 84, 97, 187, 190
water function zones **11**, 13–14, **14**, 15
Water Law 1988 (China) 43, 207
Water Law 2002 (China) 1, 9, 11–12, **11**, 15, 18, 22ff., 43ff., 63, 65, 76, 83, 84, 88, 97, 207
Water Management Act 2000 (New South Wales) 206, 209
water management stations (WMSs) 52ff.
water markets 90, 203
water metering 195–6
water permit regulation **11**, 12, 13, 23, 24–5, 47, 84
water quality maintenance 126, *127–9*
Water Quality Model method 98, *111*
water quality of major rivers *10*
water requirements outside the river course 102
water resource development (WRD) 6–7, 96, 97, 103
water resources 6–9, *8*: framework **13**; key issues 7–9, *8*
water resources allocation 2–3, 18, 21–35, 46: annual supply 26; application of management models 35; environmental flows 34; Hangjin Irrigation District (HID) 31–2; harmonious model 167–198; improvements 34–5; irrigation districts and urban water supply 25–6; Jinjiang/Jin River 30, *30*; lack of integration and consistency 33; legal framework 22–6, **27**; limited implementation 33; plan scenarios 148–50, *149*, **150**; poor specification 32–3; practical application 26–32, *29*, *30*; problems and challenges 32–3; river basins and regions 23–4; Shanghai 31; supply to water abstractors 23, 24–5; Yellow River 27–8, *29*, 31–2
Water Resources Demand Management Assistance Project (WRDMAP) 52
Water Resources Fee 43
water resources management 2, 5–19: challenges and recommendations

16–18; improving systems 17–18; legal and institutional arrangements 9–12, *10*, **11**; management approach and progress 14–16

water rights allocation 165–87, 190–91: based on current use 174, *175*, **177**; based on future demand 174–7, *176*, **177**, *178-9*, **180**; basic water allocation 171–2; current water use and future scenarios 172; ecological water allocation 172; ground and surface water allocation 172–3; harmonious model 169–71; initial principles 167–9; key issues 171–3; *see also* Shiyang River Basin case study

Water Rights Allocation Plan for the Downstream of Jinjiang 1996 30

water rights constitution 185–98: background 185–6; framework for ID water rights 190–97; Hangjin Irrigation District 186–90; initial water rights allocation 190–91; recommendations 197–8; towards an integrated framework in IDs 196–7, **197**; water certificates and tickets 191–3, *192*, *193*, **193**; water metering and monitoring systems 195–6; water user associations 194–5

water rights reform 45–7

water rights system, development of a 1–4

water sharing rules 88–9

water shortages 7, 9, 14, 205

water tickets 191–3, *193*, **193**

water use planning 32, 193–4

water user associations (WUAs) 16, 31, 32, 43, 44, 48, 52, 189ff., *193*, **193**, 194–5

Wei River *110*

WinXPRO 130

World Bank 49, 52, 62

World Health Organization (WHO) 67

World Trade Organization (WTO) 45

Wuwei Municipality 51, 53

Xia'an Reservoir **116**, 117, 145, **147**

Xian City 44

Xiangxi River 99, *111*

Xinjiang Uygur Autonomous Region 27

Yalong River 99, *109–12*

Yangtze River 6, 22, 40, 41, 42, 70, 73, 95, 101

Yangtze River Basin **7**, *8*, *10*, 66, 67, 70, 71, 99, **100**

Yangtze River Water Resources Protection Science Institute 96

Yellow River 3, 9, 12, 26ff., 42, *44*, 46, 48, 55, 85ff., 95ff., *109–12*, **186**, 187, *188*, *189*, 208, 212: environmental flows 95ff.; water resources allocation 27–8, *29*, 31–2, 194

Yellow River Basin 6, **7**, *8*, *10*, 13, 40, 43, 47, 55, 85–7, 144, *204*, 205, 212

Yellow River Conservancy Commission (YRCC) 26, 28, 32, 45, 47ff., 86, 87, 185, 187, 189, 212

Yellow River Water Allocation Plan 28, 85, *188*

Yellow River Water Resources Regulation 28

Ying River *110*, *111*

Yiwu County 13, 85, 88, 89

Yong'an Creek **116**, 117, 118, *124*, 132, 133, 145, **147**, 151, *152*, **156**, 156, **157**, 158, 161

Yongchang County 173, **174**, *175–6*, **177**, *178–9*, **180–182**

Yongding River 26, 66

Yongning River 116, **116**, 117, *124*, 145, 161

Yuecheng District 73, 64

Yuyao County 85

Zhangbei County *42*

Zhangjiakou City 54

Zhangye City 17, 49

Zhejiang Province 3, 27, 51, 62, **63**, 70, 73, 74, 85, 114ff., **114**, 144ff., **146**

Zhu Creek 42, **116**, **117**, *124*, 133, 145, **147**, 149, *149*, 151, *152*, **156**, 158, 161

Zhuxi Dam 116, 124, **125**, 133, 139, 156, **157**, 158, **158**

Zhuxi Reservoir 117–18, 145ff., **147**, 156ff., **157**, **160**

For Product Safety Concerns and Information please contact our EU
representative GPSR@taylorandfrancis.com
Taylor & Francis Verlag GmbH, Kaufingerstraße 24, 80331 München, Germany

* 9 7 8 0 4 1 5 8 5 2 0 3 6 *